Book Title: Smoke Component Yields from Bench-scale Fire Tests: 3. ISO 5660-1 / ASTM E 1354 with Enclosure and Variable Oxygen Concentration

Book Author: Nathan D. Marsh; Richard G. Gann;

Book Abstract: A standard procedure is needed for obtaining smoke toxic potency data for use in fire hazard and risk analyses. Room fire testing of finished products is impractical, directing attention to the use of apparatus that can obtain the needed data quickly and at affordable cost. This report presents examination of the fourth of a series bench-scale fire tests to produce data on the yields of toxic products in both pre-flashover and post-flashover flaming fires. The apparatus is the ISO 5660-1 / ASTM E 1354 cone calorimeter, modified to have an enclosure and a gas delivery system allowing variable oxygen concentration. The test specimens was cut from finished products that were also burned in room-scale tests: a sofa made of upholstered cushions on a steel frame, particleboard bookcases with a laminated finish, and household electric cable. Initially, the standard test procedure was followed. Subsequent variation in the procedure included reducing the supplied oxygen volume fraction to 0.18, 0.16, and 0.14, reducing the incident heat flux to 25 kW/m2, and reducing the gas flow rate by half. The yields of CO2 CO, HCl, and HCN were determined. The yields of other toxicants (NO, NO2, formaldehyde, and acrolein) were below the detection limits, but volume fractions at the detection limits were shown to be of limited toxicological importance relative to the detected toxicants. In general, performing the tests at the reduced oxygen volume fraction led to small increases on the toxic gas yields. The exceptions were an increase in the CO yield for the bookcase at 0.14 oxygen volume fraction. Reducing the incident heat flux had little effect on the toxic gas yields, other than increasing variability. Reducing the gas flow rate reduced the limits of detection by half, but also resulted in reduced gas yields at lower oxygen volume fractions. In none of the procedure variations did the CO yield approach the value of 0.2 found in real-scale post flashover fire tests.

Citation: NIST TN - 1762

Keyword: fire; fire research; smoke; room fire tests; fire toxicity; smoke toxicity

NIST Technical Note 1762

Smoke Component Yields from Bench-scale Fire Tests: 3. ISO 5660-1 / ASTM E 1354 with Enclosure and Variable Oxygen Concentration

Nathan D. Marsh
Richard G. Gann

http://dx.doi.org/10.6028/NIST.TN.1762

National Institute of
Standards and Technology
U.S. Department of Commerce

NIST Technical Note 1762

Smoke Component Yields from Bench-scale Fire Tests: 3. ISO 5660-1 / ASTM E 1354 with Enclosure and Variable Oxygen Concentration

Nathan D. Marsh
Richard G. Gann

Fire Research Division
Engineering Laboratory

http://dx.doi.org/10.6028/NIST.TN. 1762

December 2013

U.S. Department of Commerce
Penny Pritzker, Secretary

National Institute of Standards and Technology
Patrick D. Gallagher, Under Secretary of Commerce for Standards and Technology and Director

Certain commercial entities, equipment, or materials may be identified in this
document in order to describe an experimental procedure or concept adequately.
Such identification is not intended to imply recommendation or endorsement by the
National Institute of Standards and Technology, nor is it intended to imply that the
entities, materials, or equipment are necessarily the best available for the purpose.

National Institute of Standards and Technology Technical Note 1762
Natl. Inst. Stand. Technol. Tech. Note 1762, 43 pages (December 2013)
http://dx.doi.org/10.6028/NIST.TN. 1762
CODEN: NTNOEF

ABSTRACT

A standard procedure is needed for obtaining smoke toxic potency data for use in fire hazard and risk analyses. Room fire testing of finished products is impractical, directing attention to the use of apparatus that can obtain the needed data quickly and at affordable cost. This report presents examination of the fourth of a series bench-scale fire tests to produce data on the yields of toxic products in both pre-flashover and post-flashover flaming fires. The apparatus is the ISO 5660-1 / ASTM E 1354 cone calorimeter, modified to have an enclosure and a gas delivery system allowing variable oxygen concentration. The test specimens was cut from finished products that were also burned in room-scale tests: a sofa made of upholstered cushions on a steel frame, particleboard bookcases with a laminated finish, and household electric cable. Initially, the standard test procedure was followed. Subsequent variation in the procedure included reducing the supplied oxygen volume fraction to 0.18, 0.16, and 0.14, reducing the incident heat flux to 25 kW/m^2, and reducing the gas flow rate by half.

The yields of CO_2 CO, HCl, and HCN were determined. The yields of other toxicants (NO, NO_2, formaldehyde, and acrolein) were below the detection limits, but volume fractions at the detection limits were shown to be of limited toxicological importance relative to the detected toxicants. In general, performing the tests at the reduced oxygen volume fraction led to small increases on the toxic gas yields. The exceptions were an increase in the CO yield for the bookcase at 0.14 oxygen volume fraction. Reducing the incident heat flux had little effect on the toxic gas yields, other than increasing variability. Reducing the gas flow rate reduced the limits of detection by half, but also resulted in reduced gas yields at lower oxygen volume fractions. In none of the procedure variations did the CO yield approach the value of 0.2 found in real-scale postflashover fire tests.

Keywords: fire, fire research, smoke, room fire tests, fire toxicity, smoke toxicity

This page intentionally left blank

TABLE OF CONTENTS

LIST OF FIGURES

LIST OF TABLES

I. INTRODUCTION

A. CONTEXT OF THE RESEARCH

Estimation of the times that building occupants will have to escape, find a place of refuge, or survive in place in the event of a fire is a principal component in the fire hazard or risk assessment of a facility. An accurate assessment enables public officials and facility owners to provide a selected or mandated degree of fire safety with confidence. Without this confidence, regulators and/or designers tend to apply large safety factors to lengthen the tenable time. This can increase the cost in the form of additional fire protection measures and can eliminate the consideration of otherwise desirable facility designs and construction products. Error in the other direction is also risky, in that if the time estimates are incorrectly long, the consequences of a fire could be unexpectedly high.

Such fire safety assessments now rely on some form of computation that takes into account multiple, diverse factors, including the facility design, the capabilities of the occupants, the potential growth rate of a design fire, the spread rates of the heat and smoke, and the impact of the fire effluent (toxic gases, aerosols, and heat) on people who are in or moving through the fire vicinity.[1] The toolkit for these assessments, while still evolving, has achieved some degree of maturity and quality. The kit includes such tools as:

- Computer models of the movement and distribution of fire effluent throughout a facility.

  o Zone models, such as CFAST[2], have been in use for over two decades. This model takes little computational time, a benefit achieved by simplifying the air space in each room into two zones. A number of laboratory programs, validation studies,[3] and reconstructions of actual fires have given credence to the predictions.[4]

  o Computational fluid dynamics (CFD) models, such as the Fire Dynamics Simulator (FDS)[5], have seen increased use over the past decade. FDS is more computationally intense than CFAST in order to provide three-dimensional temperature and species concentration profiles. There has been extensive verification and validation of FDS predictions.[5]

  These models calculate the temperatures and combustion product concentrations as the fire develops. These profiles can be used for estimating when a person would die or be incapacitated, *i.e.*, is no longer able to effect his/her own escape.

- Devices such as the cone calorimeter[6] and larger scale apparatus[7], which are routinely used to generate information on the rate of heat release as a commercial product burns.

- A number of standards from ISO TC92 SC3 that provide support for the generation and use of fire effluent information in fire hazard and risk analyses.[8] Of particular importance is ISO 13571, which provides consensus equations for estimating the human incapacitating exposures to the narcotic gases, irritant gases, heat, and smoke generated in fires.[9]

More problematic are the sources of data for the production of the harmful products of combustion. Different materials can generate fire effluent with a wide range of toxic potencies. Most furnishing and interior finish products are composed of multiple materials assembled in a variety of geometries, and there is as of yet no methodology for predicting the evolved products from these complex assemblies. Furthermore, the generation of carbon monoxide (CO), the most common toxicant, can vary by orders of magnitudes, depending on the fire conditions.[10]

An analysis of the U.S. fire fatality data[11] showed that post-flashover fires comprise the leading scenarios for life loss from smoke inhalation. Thus, it is most important to obtain data regarding the generation of harmful species under post-flashover (or otherwise underventilated) combustion conditions. Data for pre-flashover (well-ventilated) conditions have value for ascertaining the importance of prolonged exposure to "ordinary" fire effluent and to short exposures to effluent of high potency.

B. OBTAINING INPUT DATA

The universal metric for the generation of a toxic species from a burning specimen is the yield of that gas, defined as the mass of the species generated divided by the consumed mass of the specimen.[12] If both the mass of the test specimen and the mass of the evolved species are measured continuously during a test, then it is possible to obtain the yields of the evolved species as the burning process, and any chemical change within the specimen, proceeds. If continuous measurements are not possible, there is still value in obtaining a yield for each species integrated over the burning history of the test specimen.

The concentrations of the gases (resulting from the yields and the prevalent dilution air) are combined using the equations in ISO 13571 for a base set of the most prevalent toxic species. Additional species may be needed to account for the toxic potency of the fire-generated environment.

To obtain an indicator of whether the base list of toxic species needs to be enhanced, living organisms should also be exposed to the fire effluent. The effluent exposure that generates an effect on the organisms is compared to the effect of exposure to mixtures of the principal toxic gases. Disagreement between the effluent exposure and the mixed gas exposure is an indicator of effluent components not included in the mixed gas data or the existence of synergisms or antagonisms among the effluent components. This procedure has been standardized, based on data developed using laboratory rats.[13,14] However, it is recognized that animal testing is not always possible. In these cases, it is important to identify, from the material degradation chemistry, a reasonable list of the degradation and combustion products that might be harmful to people.

Typically, the overall effluent from a harmful fire is determined by the large combustibles, such as a bed or a row of auditorium seats. The ideal fire test specimen for obtaining the yields of effluent components is the complete combustible item, with the test being conducted in an enclosure of appropriate size. Unfortunately, reliance on real-scale testing of commercial products is impractical, both for its expense per test and for the vast number of commercial products used in buildings. Such testing *is* practical for forensic investigations in which there is knowledge of the specific items that combusted.

A more feasible approach for obtaining toxic gas yields for facility design involves the use of a physical fire model – a small-scale combustor that captures the essence of the combustible and of the burning environment of interest. The test specimen is an appropriate cutting from the full combustible. To have confidence in the accuracy of the effluent yields from this physical fire model, it must be demonstrated:

- How to obtain, from the full combustible, a representative cutting that can be accommodated and burned in the physical fire model;

- That the combustion conditions in the combustor (with the test specimen in place) are related to the combustion conditions in the fire of interest, generally pre-flashover flaming (well ventilated or underventilated), post-flashover flaming, pyrolysis, or smoldering;

- How well, for a diverse set of combustible items, the yields from the small-scale combustor relate to the yields from real-scale burning of the full combustible items; and

- How sensitive the effluent yields are to the combustor conditions and to the manner in which the test specimen was obtained from the actual combustible item.

At some point, there will be sufficient data to imbue confidence that testing of further combustibles in a particular physical fire model will generate yields of effluent components with a consistent degree of accuracy.

The National Institute of Standards and Technology (NIST) has completed a project to establish a technically sound protocol for assessing the accuracy of bench-scale device(s) for use in generating fire effluent yield data for fire hazard and risk evaluation. In this protocol, the yields of harmful effluent components are determined for the real-scale burning of complete finished products during both pre-flashover and post-flashover conditions. Specimens cut from these products are then burned in various types of bench-scale combustors using their standard test protocols. The test protocols are then varied within the range of the combustion conditions related to these fire stages to determine the sensitivity of the test results to the test conditions and to provide a basis for improving the degree of agreement with the yields from the room-scale tests.

This report continues with a brief description of the previously conducted room fire tests. The full details can be found in Reference 15. Following this recap are the details of the tests using the fourth of four bench-scale apparatus to be examined.

C. PRIOR ROOM-SCALE TESTS

1. Test Configuration

With additional support from the Fire Protection Research Foundation, NIST staff conducted a series of room-fire tests of three complex products. The burn room was 2.44 m wide, 2.44 m high, and 3.66 m long (8 ft x 8 ft x 12 ft). The attached corridor was a 9.75 m (32 ft) long extension of the burn room. A doorway 0.76 m (30 in.) wide and 2.0 m (80 in.) high was centered in the common wall. The downstream end of the corridor was fully open.

2. Combustibles

Three fuels were selected for diversity of physical form, combustion behavior, and the nature and yields of toxicants produced. Supplies of each of the test fuels were stored for future use in bench-scale test method assessment.

- "Sofas" made of up to 14 upholstered cushions supported by a steel frame. The cushions consisted of a zippered cotton-polyester fabric over a block of a flexible polyurethane (FPU) foam. The fire retardant in the cushion padding contains chlorine atoms. Thus, this fuel would be a source of CO_2, CO, HCN, HCl, and partially combusted organics. The ignition source was the California TB133 propane ignition burner[16] faced downward, centered over the center of the row of seat cushions. In all but two of the tests, the sofa was centered along the rear wall of the burn room facing the doorway. In the other two tests, the sofa was placed in the middle of the room facing away from the doorway to compare the burning behavior under different air flow conditions. Two of the first group of sofa tests were conducted in a closed room to examine the effect of vitiation on fire effluent generation. In these, an electric "match" was used to initiate the fires.

- Particleboard (ground wood with a urea formaldehyde binder) bookcases with a laminated polyvinylchloride (PVC) finish. This fuel would be a source of CO_2, CO, partially combusted organics, HCN, and HCl. To sustain burning, two bookcases were placed in a "V" formation, with the TB133 burner facing upward under the lower shelves.

- Household wiring cable, consisting of two 14 gauge copper conductors insulated with a nylon and a polyvinyl chloride (PVC), an uninsulated ground conductor, two paper filler strips, and an outer jacket of a plasticized PVC. This fuel would be a source of CO_2, CO, HCl, and partially combusted organics. Two cable racks containing 3 trays each supported approximately 30 kg of cable in each of the bottom two trays and approximately 17 kg in each of the middle and top trays. The cable trays were placed parallel to the rear of the burn room. Twin propane ignition burners were centered under the bottom tray of each rack.

The elemental chemistry of each combustible was determined by an independent testing laboratory. More details regarding the elemental analysis can be found in NIST TN 1760[17]. The elemental composition of the component materials in the fuels is shown in Table 1.

Table 1. Elemental Analysis of Fuel Components.

Sample	C	H	N	Cl	P	O
Bookcase	0.481 ± 0.6 %	0.062 ± 0.8 %	0.029 ± 13 %	0.0030 ± 4 %	NA	0.426 ± 1 %
Sofa	0.545 ± 1 %	0.080 ± 1 %	0.100 ± 1 %	0.0068 ± 16 %	0.0015 ± 17 %	0.267 ± 4 %
Cable	0.576 ± 0.5 %	0.080 ± 1.5 %	0.021 ± 6 %	0.323 ± 0.4 %	NA	NA

The uncertainties are the standard deviation of three elemental analysis tests, combined with the uncertainty of the mass fraction of individual material components, i.e. fabric vs. foam for the sofa.

D. PHYSICAL FIRE MODELS

Historically, there have been numerous bench-scale devices that were intended for measuring the components of the combustion effluent.[18,19] The combustion conditions and test specimen configuration in the devices vary widely, and some devices have flexibility in setting those conditions. Currently, ISO TC92 SC3 (Fire Threat to People and the Environment) is proceeding toward standardization of one of these devices, a tube furnace (ISO/TS 19700[20]) and is considering standardization of another, the cone calorimeter (ISO 5660-1[21]) with a controlled combustion environment. There are concurrent efforts in Europe and ISO to upgrade the chemical analytical capability for a closed box test (ISO 5659-2[22]). Thus, before too long there may well be diverse (and perhaps conflicting) data on fire effluent component yields available for any given product. This situation does not support either assured fire safety or marketplace stability.

In related work, we report toxic gas measurements in the NFPA 269 toxicity test method,[17] the ISO/TS 19700 tube furnace[23], and the smoke density chamber.[24]

The modification of the cone calorimeter to measure gas yields in vitiated environments was first introduced in the early 1990s[25–27] and has recently received renewed interest.[28–30] The more recent work has focused on the effect of secondary oxidation on the heat release rate measurement, and progress on measuring chemical species is just beginning.

This page intentionally left blank

II. EXPERIMENTAL INFORMATION

A. SUMMARY OF ISO 5660-1 / ASTM E 1354 APPARATUS

1. Hardware

The standard apparatus consists of a load cell, specimen holder, truncated cone electrical resistance heater, spark igniter, canopy hood, and exhaust ductwork. The electrical heater is calibrated using a Schmidt-Boelter heat flux gauge, which itself is calibrated using a standardized source linked to a primary calibration standard. Exhaust flow in the duct is normally controlled by a variable-speed fan of the "squirrel cage" design, and measured by measuring the pressure drop across an orifice plate installed downstream from the fan, along with the temperature at that location.

Under standard operation, the heat release rate is measured via oxygen consumption calorimetry, thus the oxygen concentration in the exhaust must be measured. See the following section for more detail.

For this work, the apparatus was modified to include 1) an enclosure measuring 430 mm by 500 mm by 570 mm high, constructed of aluminum framing and polycarbonate walls (insulated in the locations of highest heat flux with aluminum foil), which seals against the underside of the canopy hood, and 2) a gas delivery system capable of delivering at 25 L/s a mixture of air and nitrogen. This consisted of a self-pressurizing 180 L liquid nitrogen dewar and a 120 psi supply of filtered "house" air. A standard CGA 580 regulator on gas outlet of the dewar maintained pressure at 80 psi, and was connected to a control manifold by 13 mm diameter copper tubing. Valves on the manifold allowed individual control of both the nitrogen gas and the compressed air, which were metered by 3 high-capacity rotameters (2 in parallel for air and one for nitrogen). The rotameters were then connected to a gas mixing chamber constructed out of 400 mm by 60 mm tube, with 2 air and 2 nitrogen inlets on one end and 4 mixed gas outlets on the other. These outlets were then each connected by more 13 mm tubing to each side of the enclosure at the base frame. The interior edges of this (hollow) aluminum frame were perforated on each side by 5 equally spaced holes, 6.4 mm diameter, providing an even supply of gas around the perimeter at the base of the enclosure. Some features of this setup can be seen in Figure 1.

One face of the enclosure includes a latchable door (20 cm square) for loading and unloading the test specimen. In order to allow time for the desired gas mixture to expel air introduced during specimen loading, a shutter composed of ceramic board supported by a metal frame and slide mechanism was placed between the specimen and the cone heater. Attached to the shutter was a rod extending through a bulkhead fitting in the enclosure wall so that the shutter could be withdrawn to commence a test. During operation, balancing of the gas supply and the exhaust was accomplished by observing a small perforation in the enclosure provided with soap film, while monitoring the recorded volumetric flow rate calculated from the pressure differential across the orifice plate in the exhaust duct.

Figure 1. Modified Cone Calorimeter

Although not discussed here, the heat release rate measurement was calibrated daily using a 5 kW methane burner, itself calibrated by measuring the gas flow using a piston-type primary standard calibrator, accurate to 1 % of the reading.

2. Gas Sampling and Analysis Systems

In the room-scale tests (Section I.C), measurements were made of 12 gases. Water and methane were included because of their potential interference with the quantification of the toxic gases. Two of the toxic gases, HBr and HF, were not found in the combustion products because there was no fluorine or bromine in the test specimens. The remaining eight toxic gases were acrolein (C_3H_4O), Carbon monoxide (CO), carbon dioxide (CO_2), formaldehyde (CH_2O), hydrogen chloride (HCl), hydrogen cyanide (HCN), nitric oxide (NO), and nitrogen dioxide (NO_2). Some of these turned out to be generated at levels that would not have contributed significantly to the incapacitation of exposed people. Thus, it was deemed unlikely that animal tests would have added much tenability information. As a result, the same gases were monitored in the bench-scale tests, and no animals were exposed. The basis for comparison between tests of the same combustibles at the two scales is the yields of the chemically diverse set of toxicants.

CO and CO_2 were quantified using a nondispersive infrared (NDIR) gas analyzer; oxygen was quantified by a paramagnetic analyzer in the same instrument. The precisions of the analyzers, as provided by the manufacturer, were:

> CO: 10 µL/L
>
> CO_2: 0.02 L/L
>
> O_2: 0.05 L/L

Sampling for this instrument was pulled from the exhaust duct upstream from the exhaust fan, using a perforated ring sampling probe. The flow passed through a large volume cartridge filter, through two parallel membrane HEPA filters, through an electrical chiller maintained at -2 °C to -5 °C, and finally through a fixed bed of calcium sulfate desiccant. The stream sent to the O_2 analyzer also passed through a fixed bed of sodium hydroxide-coated silica particles. The flows were maintained at 3 L/min. The analyzer itself was calibrated daily with zero and span gases (a mixture of 2800 µL/L CO and 0.028 L/L of CO_2 in nitrogen, and ambient air (0.2095 L/L oxygen on a dry basis)). The span gas is certified to be accurate to within 2 % of the value.

The concentrations of CO and the additional six toxic gases were measured using a Bruker Fourier transform infrared (FTIR) spectrometer[*] equipped with an electroless nickel plated aluminum flow cell (2 mm thick KBr windows and a 1 m optical pathlength) with an internal volume of 0.2 L, maintained at (170 ± 5) °C. Samples were drawn through a heated 6.35 mm (¼ in.) stainless steel tube from inside the exhaust duct, upstream of the exhaust, with its tip at approximately the centerline. The sample was pulled through the sampling line and flow cell by a small pump located downstream from the flow cell. There were no traps or filters in this sampling line. The pump flow was measured at 4 L/min maximum, but was at times lower due to fouling of the sampling lines with smoke deposits.

Although this instrument has a longer path length and is therefore more sensitive than the one used in our previous studies, the "batch" nature of a cone calorimeter experiment along with the continuous flow of the exhaust requires instantaneous data collection, in this case at 0.17 Hz, that did not allow for data averaging to improve the signal-to-noise. Therefore, the limits of detection are essentially unchanged from the previous studies. The implications of this will be discussed in Section V.

An example of a spectrum measured by FTIR spectroscopy during one such test is displayed in Figure 2. The series of peaks extending from about 3050 cm^{-1} to 2600 cm^{-1} are due to HCl. In this case, it is possible to resolve the individual frequencies corresponding to changes in the population of rotational states as the H-Cl bonds vibrate. This is usually only possible for small gas phase molecules. There are three spectral features due to CO_2 that are evident in this spectrum. The most intense, centered at 2350 cm^{-1}, corresponds to asymmetric stretching of the two C=O bonds. The symmetric stretch is not observed because there is no change in dipole moment when both O atoms move in phase. The second feature, seen as two distinct peaks centered at about 3650 cm^{-1}, is an overtone band that derives from the simultaneous excitation of these bond-stretching modes. The third peak at about 650 cm^{-1} is due to the out of plane bending of the molecule. There are bands due to the C≡O stretching vibrations in carbon monoxide, centered at about 2150 cm^{-1}. The remaining peaks in this spectrum are due to H_2O.

---

[*] Certain commercial equipment, products, or materials are identified in this document in order to describe a procedure or concept adequately or to trace the history of the procedures and practices used. Such identification is not intended to imply recommendation, endorsement, or implication that the products, materials or equipment are necessarily the best available for the purpose.

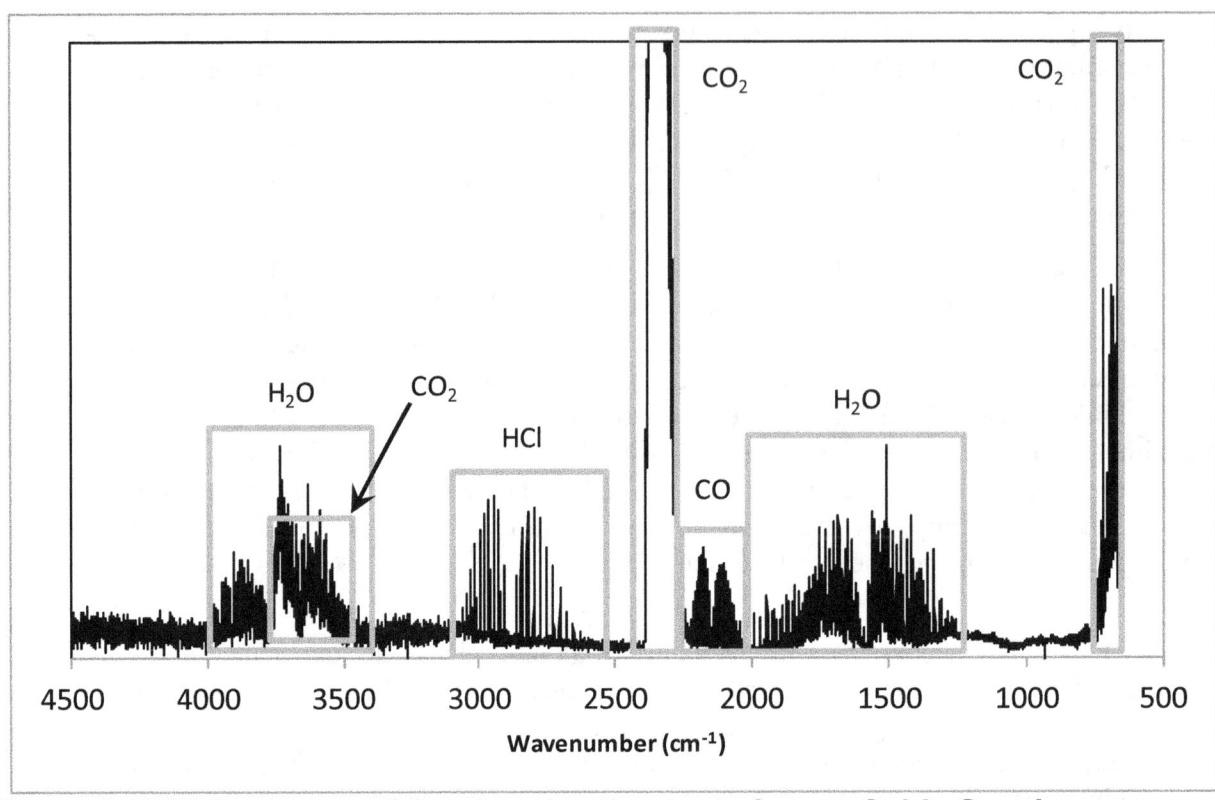

Figure 2. FTIR Spectrum of Combustion Products from a Cable Specimen

Using these spectra, gas concentrations were quantified using the Autoquant software. This is a software package for performing real time and off-line quantitative analyses of target compounds, and is based on the Classical Least Squares (CLS) algorithm as described by Haaland *et al.*[31] In this method, the measured spectra are fit to linear combinations of reference spectra corresponding to the target compounds.

Calibration spectra were obtained from a quantitative spectral library assembled by Midac[32] and from a collection of spectra provided the Federal Aviation Administration who performed bench-scale fire tests on similar materials.[33] In this analysis, the least squares fits were restricted to characteristic frequency regions or windows for each compound that were selected in such a way as to maximize the discrimination of the compounds of interest from other components present in fire gases. All reference spectra were recorded at 170 °C and ambient pressure.

The identities of the target compounds (as well as other compounds that absorb at the same frequencies and must, therefore, be included in the analyses), their corresponding concentrations (expressed in units of $\mu L/L$ for a mixture of the calibration gas and N_2 in a 1 meter cell), and the characteristic spectral windows used in the quantitative analyses are listed in Table 2.

Also listed in this Table are minimum detection limits (MDLs) for each of the target compounds. These values, which represent the lowest concentrations that can be measured with the instrumentation employed in these tests, were estimated as follows. The calibration spectra were added to test spectra (which, when possible, were selected in such a way that only the compound of interest was not present) with varying coefficients until the characteristic peaks of the target compounds were just discernible above the baseline noise. The MDL values reported in Table 2 were obtained by multiplying these coefficients by the known concentrations of the target compound in the calibration mixtures.

Water, methane and acetylene are included in the quantitative analyses because they have spectral features that interfere with the target compounds. The nitrogen oxides absorb in the middle of the water band that extends from about 1200 cm^{-1} to 2050 cm^{-1}. Consequently, the real limits of detection for these two compounds are an order of magnitude higher than for any of the other target compounds. Thus, it is not surprising that their presence was not detected in any of the tests.

Table 2. Species and Frequency Windows for FTIR Analysis.

Compound	Reference Volume Fraction (µL/L)	Frequency Window (cm^{-1})	Minimum Detection Limit (µL/L)
CH_4	483	2800 to 3215	20
C_3H_4O	2250	850 to 1200	20
CH_2O	11300	2725 to 3000	40
CO_2	47,850	660 to 725, 2230 to 2300	800[a]
CO	2410	2050 to 2225	20
H_2O	100,000	1225 to 2050, 3400 to 4000	130[a]
HCl	9870	2600 to 3100	20
HCN	507	710 to 722, 3200 to 3310	35
NO	512	1870 to 1950	70
NO_2	70	1550 to 1620	40

[a] Present in the background.

Delay times for gas flows from the sampling locations within the test structure to the gas analyzers were small compared to the duration of the specimen burning. The burn durations were near 20 min for the bookcase specimens, 1 to 3 min for the sofa specimens, and 2 to 4 min for the cable specimens. Combining the gas sample pumping rate and the volumes of the sampling lines, the delay time to the oxygen analyzer was about 5 s, about 1 s to the CO and CO_2 analyzers, and 1 s to the FTIR analyzer. These delay times are long enough to allow for a small degree of axial diffusion. However, since our analysis integrates the data over time, this did not adversely affect the quantification of total gas evolved.

B. OPERATING PROCEDURES

1. "Standard" Testing

The intent was to test specimens of each of the three types under normal and reduced-oxygen conditions, two incident heat fluxes, and two gas flow rates. The steps in the procedure are:

- Calibrate the heat flux to the specimen surface and calibrate the gas analyzers (each performed daily).

- Establish the desired gas flow rates and oxygen concentration, with an empty specimen holder in place.

- Turn on the gas sampling data collection and establish background data for the combustion product concentrations.

- Open the chamber door.

- Replace the empty specimen holder with one containing the specimen.

- Close the chamber door.

- Wait for the oxygen concentration to return to its established value.

- Withdraw the shutter and insert the spark igniter.

- Turn on the spark igniter.

- Record the time of ignition; turn off and withdraw the spark igniter.

- Collect concentration data until a steady state of pyrolysis is reached.

- Open the chamber door; remove the specimen; close the chamber door.

- Record a post-test background to account for any drift.

- Weigh any specimen residue.

As in the previous work, the bookcase material underwent considerable pyrolyzing after the flames disappeared. However, as the cone calorimeter is a continuous-flow apparatus, no physical steps were necessary to isolate the pyrolysis results from the flaming results, and the transition between the two is clearly visible in the data, e.g. the differential of the mass loss data.

2. Test Specimens

The specimens were intended to approximate the full item. Specimen size was mostly limited by the size of the specimen holder, although in the case of the cables it was not necessary to completely fill the pan or use multiple layers. Instead the number of cables was chosen to produce similar gas concentrations and heat release rates compared to the other two types of specimens. The bookcase specimens were single, 10 cm x 10 cm pieces of the particle board, with the vinyl surface facing up. The sofa specimens were each a single piece of foam, 10 cm by 10 cm by 1 cm thick, covered with a single piece of the polyester/cotton cover fabric, 10.5 cm x 10.5 cm, on the upper (exposed) surface (the fabric being slightly oversized so that it could be "tucked" in around the edges of the foam). The cable specimens were five 10 cm lengths of cable cut from the spool and placed side-by-side. All specimens were wrapped on the bottom and sides with aluminum foil, which was cut to be flush with the top surface of the specimen. The specimen holder was lined with a refractory aluminosilicate blanket, adjusted in thickness so that each specimen was the same distance from the upper surface to the cone heater assembly (28 mm). Note that with the exception of area dimensions, these specimens were prepared almost identically to those in Ref. 17, and are therefore not reproduced here. Also, in our prior work we found little to no impact from "dicing" the specimens, i.e. cutting them into small pieces, and therefore did not investigate that effect here.

3. Test Procedure Variation

One of the purposes of this program was to obtain effluent composition data in tests with variants on the standard operating procedure. This would enable examination of the potential for an improved relationship with the yield data from the room-scale tests, as well as an indication of the sensitivity of the gas yields to the specified operating conditions

- Variation in incident heat flux between 20 kW/m^2 and 50 kW/m^2, to determine the significance on evolved gases.

- Variation of the available oxygen volume fraction from 21 % to 14 %, to more closely approach the post-flashover conditions that occurred in the room-scale tests. During pre-flashover burning, the air entrained by a fire has an oxygen volume fraction of nominally 0.21. This fraction is lower for post-flashover fires. As part of this work it was determined that combustion was barely sustainable at an oxygen volume fraction of 0.14.

- Variation of overall gas flow rate from 25 L/s to 12.5 L/s in order to increase sensitivity in gas measurements, and to allow the reduction of oxygen volume fraction to 0.14, which was difficult in our system at the higher flow. A known drawback to reducing the flow is that the sensitivity of the flow measurement, via the pressure drop across the orifice plate, is reduced. (This can be ameliorated by installing a smaller orifice so that the pressure drop in brought back up to the optimal range of the pressure sensor.)

C. DATA COLLECTION

The signals from the load cell, temperature, and pressure needed for ISO 5660-1, and from the fixed gas analyzer were recorded on a personal computer using a custom-made data acquisition system based on National Instruments data acquisition hardware. Values were recorded at 1 s intervals. The FTIR spectra were recorded using the software package provided by the manufacturer. Spectra were recorded every 6 s.

This page intentionally left blank

III. CALCULATION METHODS

A. MASS LOSS RATE

The specimen mass loss during a test was determined from the initial reading from the load cell and from a point where the specimen transitioned from burning to pyrolyzing, which was characterized by a sudden drop in the mass loss rate (derivative of the mass loss calculated over 16 data points / seconds). The uncertainty in the mass loss, derived from the uncertainties in these two measurements, was 0.1 g.

B. NOTIONAL GAS YIELDS

The notional, or maximum possible, gas yields (Table 3) were calculated as follows:

- CO_2: Assume all the carbon in the test specimen is converted to CO_2. Multiply the mass fraction of C in the test specimen (Table 1) by the ratio of the molecular mass of CO_2 to the atomic mass of carbon.

- CO: Assume all the carbon in the test specimen is converted to CO. Multiply the mass fraction of C by the ratio of the molecular mass of CO to the atomic mass of carbon.

- HCN: Assume all the nitrogen in the test specimen is converted to HCN. Multiply the mass fraction of N by the ratio of the molecular mass of HCN to the atomic mass of nitrogen.

- HCl: Assume all the chlorine in the test specimen is converted to HCl. Multiply the mass fraction of Cl by the ratio of the molecular mass of HCl to the atomic mass of chlorine.

The notional yields from the bookcase and cable specimens were assumed to be the same as the yields from the intact combustibles.[15] The sofa specimen had a mass ratio of fabric to foam that differed modestly from the intact sofas.

Table 3. Calculated Notional Yields of Toxic Products from the Test Specimens.

Gas	Notional Yields		
	Bookcase	Cable	Sofa
CO_2	1.72 ± 1 %	2.11 ± 1 %	1.95 ± 4 %
CO	1.09 ± 1 %	1.33 ± 1 %	1.24 ± 4 %
HCN	0.057 ± 13 %	0.040 ± 6 %	0.193 ± 4 %
HCl	0.0026 ± 4 %	0.332 ± 1 %	0.0069 ± 19 %

The uncertainty in the notional yield values is determined by the uncertainty in the prevalence of the central element (in the bullets just above) in the combustible. For the cuttings from the sofas, the uncertainty in the notional yields was increased by the small variability (estimated at 3 percent) in the relative masses of the fabric and padding materials in the test specimens.

C. CALCULATED GAS YIELDS

1. CO and CO_2

Yields of CO and CO_2 were calculated using the measured gas concentrations from the NDIR instrument, the measured flows in the exhaust, the exhaust temperature, (accounting for the background concentrations in room air), the consumed mass of the fuel, and the ideal gas law. Concentrations were converted to mass flow, then integrated numerically (rectangular) over the time of flaming combustion.

As we observed previously,[17] the CO_2 absorption band in the FTIR is saturated at normal volume fractions and is therefore highly non-linear. As both the FTIR and NDIR measurements were taken from the same location, it was not necessary to measure the CO_2 concentration by FTIR.

Figure 3 shows a comparison of the maximum CO volume fractions measured by the two instruments. The FTIR consistently found a lower volume fraction than the NDIR above 300 μL/L. This is likely due to the non-linearity inherent in IR spectra at higher absorption. Since the spectrometer used in this study had a path length that was 10 times that used in the previous work,[17,23,24] it is not unreasonable that these data fall more in the non-linear range. Therefore, the data reported here for CO are taken from the NDIR results.

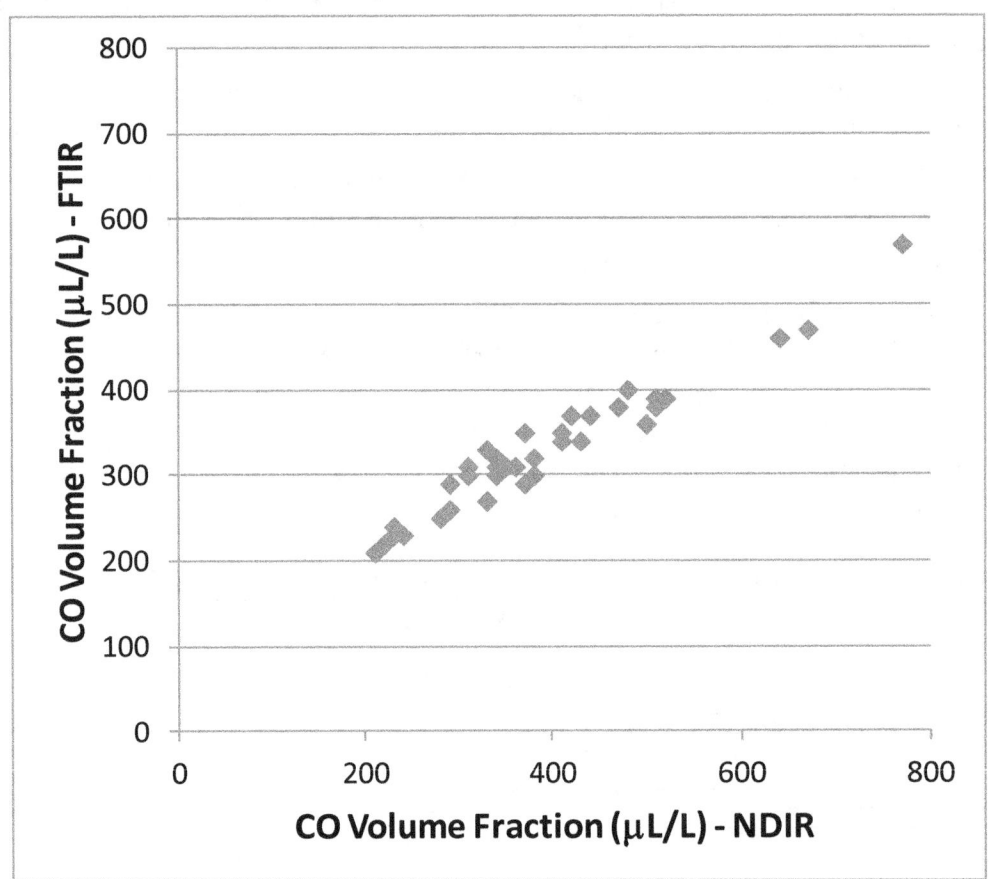

Figure 3. Comparison of FTIR and NDIR Measurements of CO Volume Fraction

2. HCl and HCN

The only calculable HCN yields were from the sofa specimens. The only calculable HCl yields were from the cable specimens. FTIR spectra from these experiments were analyzed as described in section II.A.2. This analysis normally includes an uncertainty of 10 % of the reported value. Volume fractions were converted into mass flows using the flow and temperature data, then numerically integrated over the period of flaming combustion. Because these data are more sparse than those from the NDIR (0.17 Hz vs. 1.0 Hz) a trapezoidal integration was used. We considered the possibility that these results suffered from the same non-linear relation between absorption and concentration that occurred with the CO measurements. However, since the HCN concentrations were quite low, and the HCl yields were reasonably close to their notional yields (see Table 11) and similar to the previous work using a different instrument with a shorter path length,[17,23,24] we determined that non-linear effects were not significant. Furthermore, because the lower heat flux did not, by any other indication, have a significant impact on gas yields, other than increased variability, the quantification of FTIR data from these runs was omitted.

3. Other Gases

The volume fractions of the other toxic gases were always below the detection limit. Thus, the upper limits of the yields of these gases were estimated using their limits of detection.

This page intentionally left blank

IV. RESULTS

A. TESTS PERFORMED

The following is the test numbering key, with format F(2)-q-[O₂]-N, where

F:	Fuel [S = sofas; B = bookcases; C = cable]
(2):	present for the 12.5 L/s flow rate, absent for the 25 L/s flow rate
q:	heat flux per unit area (kW/m²)
[O₂]	Approximate initial oxygen volume percent in the supply gas
N:	Replicate test number for that set of combustible and conditions

Table 4 through Table 9 present the test data and the calculated yields for the bookcase, sofa, and cable specimens, respectively. In these tables, Δt is the observed duration of flaming combustion and Δm is the measured mass lost during the flaming combustion. Volume fractions represent the maximum value in the test, usually soon after ignition. A yield number in red indicates a potential outlier that, if discarded, could improve the repeatability under those conditions.

The horizontal shaded bands highlight groups of replicate tests.

Table 4. Data from Bookcase Material Tests

Code	Specimen Mass (g)	Δm (g)	Δt (s) Flaming	NDIR F_{CO_2} (L/L)	F_{CO} (μL/L)	F_{O_2} (L/L)
b-50-21-1	149.8	106.0	1068	0.0065	120	0.2037
b-50-21-2	135.6	105.2	1090	0.0067	140	0.2033
b-50-21-3	138.4	103.7	1043	0.0072	150	0.2028
b-50-18-1	136.0	103.8	1110	0.0070	160	0.1731
b-50-18-2	137.4	101.5	1060	0.0070	160	0.1730
b-50-18-3	138.1	102.1	1110	0.0064	160	0.1748
b-50-16-1	119.2	98.4	993	0.0066	140	0.1559
b-50-16-2	136.6	101.7	1115	0.0062	170	0.1545
b-50-16-3	120.6	91.6	940	0.0062	150	0.1548
b2-50-21-1	131.5	99.6	1110	0.0121	240	0.1974
b2-50-21-2	130.2	97.3	1010	0.0123	230	0.1972
b2-50-21-3	127.4	96.9	1040	0.0124	230	0.1970
b2-50-16-1	131.6	99.7	1140	0.0106	250	0.1504
b2-50-16-2	128.2	97.3	1050	0.0103	230	0.1495
b2-50-16-3	123.8	93.8	1070	0.0103	230	0.1499
b2-50-14-1	129.7	95.7	1170	0.0099	850	0.1285
b2-50-14-2	132.2	98.1	1160	0.0100	450	0.1297
b2-50-14-3	131.6	98.2	1200	0.0101	460	0.1298
b-25-21-1	124.6	93.3	1295	0.0053	180	0.2051
b-25-21-3	128.6	103.0	1590	0.0057	200	0.2046
b-25-21-4	138.8	102.5	1590	0.0058	190	0.2045
b-25-18-1	126.6	34.6	590	0.0045	220	0.1675
b-25-18-2	127.4	30.3	540	0.0043	280	0.1773
b-25-18-3	130.9	54.5	620	0.0042	200	0.1593
b-25-18-4	122.5	32.2	550	0.0041	190	0.1769
b-25-16-1	128.4	26.9	620	0.0037	210	0.1578
b-25-16-2	129.6	21.9	450	0.0037	230	0.1594
b-25-16-3	128.6	25.8	520	0.0032	220	0.1590

Table 5. Data from Sofa Material Tests

Code	Specimen Mass (g)	Δm (g)	Δt (s) Flaming	NDIR			FTIR	
				F_{CO_2} (L/L)	F_{CO} (μL/L)	F_{O_2} (L/L)	F_{CO} (μL/L)	F_{HCN} (μL/L)
s-50-21-1	12.0	8.5	62	0.0071	290	0.2012	260	80
s-50-21-2	11.8	8.5	65	0.0074	310	0.2007	310	90
s-50-21-3	11.7	8.7	66	0.0076	310	0.2007	300	90
s-50-18-1	12.4	8.6	82	0.0067	370	0.1727	290	150
s-50-18-2	11.8	8.5	125	0.0064	330	0.1730	270	130
s-50-18-3	11.9	8.6	105	0.0066	380	0.1727	300	170
s-50-16-1	12.2	8.7	115	0.0068	470	0.1522	380	270
s-50-16-2	12.1	8.4	110	0.0067	420	0.1524	370	250
s-50-16-3	12.3	9.0	120	0.0071	430	0.1527	340	230
s2-50-21-1	11.5	8.5	85	0.0149	520	0.1903	390	210
s2-50-21-2	11.8	8.7	100	0.0120	480	0.1944	400	180
s2-50-21-3	11.8	8.7	90	0.0138	510	0.1919	390	170
s2-50-16-1	10.6	7.7	100	0.0119	500	0.1445	360	330
s2-50-16-2	11.7	8.6	130	0.0102	640	0.1485	460	380
s2-50-16-3	11.7	8.6	120	0.0106	770	0.1481	570	460
s2-50-14-1	11.2	8.0	120	0.0105	440	0.1265	370	110
s2-50-14-2	11.7	8.5	120	0.0104	510	0.1266	380	180
s2-50-14-3	11.9	8.5	120	0.0115	670	0.1252	470	250
s-25-21-1	12.4	8.5	150	0.0059	170	0.2041		
s-25-21-2	12.1	6.4	115	0.0054	160	0.2045		
s-25-21-3	12.2	8.1	170	0.0055	170	0.2044		
s-25-18-1	12.2	5.3	165	0.0043	130	0.1758		
s-25-18-2	12.2	5.5	150	0.0045	130	0.1756		
s-25-18-3	12.3	5.0	110	0.0049	180	0.1755		
s-25-18-4	12.2	9.1	170	0.0044	140	0.1748		
s-25-18-5	12.2	5.6	120	0.0047	170	0.1751		
s-25-16-1	12.2	3.5	135	0.0007	50	0.1601		
s-25-16-3	12.3	5.1	105	0.0039	110	0.1593		
s-25-16-4	12.2	3.5	180	0.0008	50	0.1361		

Table 6. Data from Cable Material Tests

Code	Specimen Mass (g)	Δm (g)	Δt (s) Flaming	NDIR			FTIR	
				F_{CO_2} (L/L)	F_{CO} (μL/L)	F_{O_2} (L/L)	F_{CO} (μL/L)	F_{HCl} (μL/L)
c-50-21-1	56.9	13.8	175	0.0041	340	0.2051	310	900
c-50-21-2	38.2	9.5	169	0.0033	220	0.2051	220	810
c-50-21-3	37.7	9.1	145	0.0033	240	0.2062	230	800
c-50-18-1	35.7	8.7	180	0.0029	210	0.1779	210	750
c-50-18-2	35.6	8.7	190	0.0028	220	0.1769	220	730
c-50-18-3	36.0	8.9	175	0.0032	280	0.1764	250	820
c-50-16-1	36.2	8.7	160	0.0029	230	0.1597	230	750
c-50-16-2	35.0	7.8	135	0.0033	340	0.1562	300	880
c-50-16-3	35.2	8.1	158	0.0030	230	0.1564	240	850
c2-50-21-1	36.5	8.1	140	0.0048	330	0.2044	330	1220
c2-50-21-2	37.3	8.2	140	0.0050	340	0.2042	320	1180
c2-50-21-3	37.2	8.4	135	0.0052	370	0.2039	350	1190
c2-50-16-1	37.2	8.8	180	0.0046	410	0.1554	350	1080
c2-50-16-2	38.0	8.9	190	0.0044	350	0.1554	310	1100
c2-50-16-3	37.5	8.7	175	0.0047	410	0.1546	340	1160
c2-50-14-2	36.8	8.4	185	0.0041	380	0.1361	320	1370
c2-50-14-3	38.7	9.6	240	0.0036	290	0.1335	290	1210
c2-50-14-4	36.9	8.1	175	0.0039	360	0.1366	310	1240
c-25-21-1	38.0	8.5	220	0.0028	160	0.2072		
c-25-21-2	37.4	8.3	230	0.0027	140	0.2075		
c-25-21-3	37.2	8.1	230	0.0026	130	0.2072		
c-25-18-1	36.0	7.2	230	0.0020	120	0.1789		
c-25-18-2	35.4	6.7	260	0.0020	120	0.1786		
c-25-18-3	35.7	6.8	250	0.0019	110	0.1775		
c-25-16-1	35.7	4.8	265	0.0014	80	0.1584		
c-25-16-2	36.4	4.2	260	0.0014	80	0.1586		
c-25-16-3	35.8	4.4	245	0.0015	80	0.1604		

Table 7. Gas Yields from Bookcase Material Tests

Code	Specimen Mass (g)	Δm (g)	y_{CO2} (g/g)	y_{CO} (g/g)	y_{HCN} (g/g)	y_{HCl} (g/g)	y_{NO} (g/g)	y_{NO2} (g/g)	$y_{acrolein}$ (g/g)	y_{form} (g/g)
b-50-21-1	149.8	106.0	1.10	0.010	< 0.0084	< 0.0065	< 0.019	< 0.016	< 0.010	< 0.011
b-50-21-2	135.6	105.2	1.11	0.010	< 0.0086	< 0.0066	< 0.019	< 0.017	< 0.010	< 0.011
b-50-21-3	138.4	103.7	1.13	0.009	< 0.0083	< 0.0064	< 0.018	< 0.016	< 0.010	< 0.011
b-50-18-1	136.0	103.8	1.09	0.013	< 0.0087	< 0.0067	< 0.019	< 0.017	< 0.010	< 0.011
b-50-18-2	137.4	101.5	1.10	0.012	< 0.0084	< 0.0065	< 0.019	< 0.016	< 0.010	< 0.011
b-50-18-3	138.1	102.1	1.10	0.013	< 0.0087	< 0.0067	< 0.019	< 0.017	< 0.010	< 0.011
b-50-16-1	119.2	98.4	1.02	0.020	< 0.0079	< 0.0061	< 0.018	< 0.015	< 0.009	< 0.010
b-50-16-2	136.6	101.7	1.05	0.020	< 0.0086	< 0.0066	< 0.019	< 0.017	< 0.010	< 0.011
b-50-16-3	120.6	91.6	1.10	0.018	< 0.0080	< 0.0062	< 0.018	< 0.016	< 0.010	< 0.010
b2-50-21-1	131.5	99.6	1.02	0.006	< 0.0045	< 0.0035	< 0.010	< 0.009	< 0.005	< 0.006
b2-50-21-2	130.2	97.3	1.01	0.006	< 0.0042	< 0.0032	< 0.009	< 0.008	< 0.005	< 0.005
b2-50-21-3	127.4	96.9	1.03	0.006	< 0.0043	< 0.0033	< 0.010	< 0.008	< 0.005	< 0.005
b2-50-16-1	131.6	99.7	1.03	0.010	< 0.0048	< 0.0037	< 0.011	< 0.009	< 0.006	< 0.006
b2-50-16-2	128.2	97.3	1.02	0.008	< 0.0045	< 0.0035	< 0.010	< 0.009	< 0.005	< 0.006
b2-50-16-3	123.8	93.8	1.03	0.009	< 0.0048	< 0.0037	< 0.011	< 0.009	< 0.006	< 0.006
b2-50-14-1	129.7	95.7	0.69	0.058	< 0.0051	< 0.0039	< 0.011	< 0.010	< 0.006	< 0.006
b2-50-14-2	132.2	98.1	0.93	0.032	< 0.0048	< 0.0037	< 0.011	< 0.009	< 0.006	< 0.006
b2-50-14-3	131.6	98.2	0.93	0.032	< 0.0049	< 0.0038	< 0.011	< 0.010	< 0.006	< 0.006
b-25-21-1	124.6	93.3	1.15	0.015						
b-25-21-3	128.6	103.0	1.12	0.020						
b-25-21-4	138.8	102.5	1.14	0.022						
b-25-18-1	126.6	34.6	0.85	0.028						
b-25-18-2	127.4	30.3	0.92	0.023						
b-25-18-3	130.9	54.5	0.55	0.018						
b-25-18-4	122.5	32.2	0.92	0.022						
b-25-16-1	128.4	26.9	0.61	0.037						
b-25-16-2	129.6	21.9	0.84	0.027						
b-25-16-3	128.6	25.8	0.86	0.027						

Table 8. Gas Yields from Sofa Material Tests

Code	Specimen Mass (g)	Δm (g)	y_{CO2} (g/g)	y_{CO} (g/g)	y_{HCN} (g/g)	y_{HCl} (g/g)	y_{NO} (g/g)	y_{NO2} (g/g)	$y_{acrolein}$ (g/g)	y_{form} (g/g)
s-50-21-1	12.0	8.5	1.39	0.026	0.0033	< 0.0055	< 0.016	< 0.014	< 0.0085	< 0.0091
s-50-21-2	11.8	8.5	1.39	0.027	0.0045	< 0.0051	< 0.015	< 0.013	< 0.0078	< 0.0084
s-50-21-3	11.7	8.7	1.45	0.027	0.0032	< 0.0054	< 0.016	< 0.014	< 0.0083	< 0.0089
s-50-18-1	12.4	8.6	1.43	0.034	0.0071	< 0.0058	< 0.017	< 0.015	< 0.0089	< 0.0096
s-50-18-2	11.8	8.5	1.48	0.036	0.0071	< 0.0054	< 0.016	< 0.014	< 0.0083	< 0.0089
s-50-18-3	11.9	8.6	1.45	0.036	0.0088	< 0.0054	< 0.016	< 0.014	< 0.0083	< 0.0089
s-50-16-1	12.2	8.7	1.38	0.045	0.0124	< 0.0056	< 0.016	< 0.014	< 0.0086	< 0.0092
s-50-16-2	12.1	8.4	1.44	0.044	0.0120	< 0.0059	< 0.017	< 0.015	< 0.0091	< 0.0097
s-50-16-3	12.3	9.0	1.41	0.044	0.0132	< 0.0060	< 0.017	< 0.015	< 0.0092	< 0.0099
s2-50-21-1	11.5	8.5	1.34	0.024	0.0035	< 0.0025	< 0.007	< 0.006	< 0.0038	< 0.0041
s2-50-21-2	11.8	8.7	1.40	0.024	0.0036	< 0.0029	< 0.008	< 0.007	< 0.0045	< 0.0048
s2-50-21-3	11.8	8.7	1.34	0.025	0.0038	< 0.0027	< 0.008	< 0.007	< 0.0042	< 0.0044
s2-50-16-1	10.6	7.7	1.31	0.032	0.0059	< 0.0033	< 0.010	< 0.008	< 0.0051	< 0.0054
s2-50-16-2	11.7	8.6	1.42	0.036	0.0114	< 0.0032	< 0.009	< 0.008	< 0.0049	< 0.0053
s2-50-16-3	11.7	8.6	1.40	0.037	0.0114	< 0.0030	< 0.009	< 0.008	< 0.0046	< 0.0049
s2-50-14-1	11.2	8.0	1.38	0.033	0.0022	< 0.0036	< 0.010	< 0.009	< 0.0055	< 0.0059
s2-50-14-2	11.7	8.5	1.41	0.032	0.0041	< 0.0030	< 0.009	< 0.008	< 0.0046	< 0.0049
s2-50-14-3	11.9	8.5	1.38	0.034	0.0054	< 0.0032	< 0.009	< 0.008	< 0.0049	< 0.0053
s-25-21-1	12.4	8.5	1.56	0.024						
s-25-21-2	12.1	6.4	1.45	0.024						
s-25-21-3	12.2	8.1	1.63	0.024						
s-25-18-1	12.2	5.3	0.91	0.030						
s-25-18-2	12.2	5.5	1.06	0.028						
s-25-18-3	12.3	5.0	1.32	0.028						
s-25-18-4	12.2	9.1	1.42	0.024						
s-25-18-5	12.2	5.6	1.94	0.037						
s-25-16-1	12.2	3.5	0.15	0.022						
s-25-16-3	12.3	5.1	1.36	0.029						
s-25-16-4	12.2	3.5	0.16	0.030						

Table 9. Gas Yields from Cable Material Tests

Code	Specimen Mass (g)	Δm (g)	y_{CO2} (g/g)	y_{CO} (g/g)	y_{HCN} (g/g)	y_{HCl} (g/g)	y_{NO} (g/g)	y_{NO2} (g/g)	$y_{acrolein}$ (g/g)	y_{form} (g/g)
c-50-21-1	56.9	13.8	0.94	0.056	< 0.011	0.23	< 0.024	< 0.021	< 0.013	< 0.014
c-50-21-2	38.2	9.5	1.12	0.051	< 0.016	0.22	< 0.036	< 0.031	< 0.019	< 0.020
c-50-21-3	37.7	9.1	1.06	0.055	< 0.014	0.24	< 0.031	< 0.027	< 0.017	< 0.018
c-50-18-1	35.7	8.7	1.16	0.066	< 0.016	0.26	< 0.036	< 0.031	< 0.019	< 0.020
c-50-18-2	35.6	8.7	1.21	0.065	< 0.017	0.26	< 0.038	< 0.033	< 0.020	< 0.022
c-50-18-3	36.0	8.9	1.14	0.071	< 0.016	0.25	< 0.036	< 0.031	< 0.019	< 0.020
c-50-16-1	36.2	8.7	1.02	0.063	< 0.017	0.29	< 0.038	< 0.033	< 0.020	< 0.022
c-50-16-2	35.0	7.8	1.15	0.078	< 0.015	0.30	< 0.033	< 0.029	< 0.018	< 0.019
c-50-16-3	35.2	8.1	1.16	0.073	< 0.020	0.32	< 0.044	< 0.039	< 0.024	< 0.025
c2-50-21-1	36.5	8.1	1.03	0.049	< 0.007	0.22	< 0.016	< 0.014	< 0.008	< 0.009
c2-50-21-2	37.3	8.2	1.04	0.049	< 0.007	0.23	< 0.016	< 0.014	< 0.008	< 0.009
c2-50-21-3	37.2	8.4	0.99	0.051	< 0.007	0.21	< 0.016	< 0.014	< 0.008	< 0.009
c2-50-16-1	37.2	8.8	0.99	0.062	< 0.008	0.23	< 0.018	< 0.016	< 0.009	< 0.010
c2-50-16-2	38.0	8.9	1.03	0.059	< 0.009	0.24	< 0.020	< 0.018	< 0.011	< 0.011
c2-50-16-3	37.5	8.7	1.04	0.061	< 0.008	0.23	< 0.018	< 0.016	< 0.009	< 0.010
c2-50-14-4	36.8	8.4	0.91	0.057	< 0.009	0.28	< 0.020	< 0.018	< 0.011	< 0.011
c2-50-14-2	38.7	9.6	0.82	0.048	< 0.009	0.28	< 0.020	< 0.018	< 0.011	< 0.011
c2-50-14-3	36.9	8.1	0.91	0.057	< 0.009	0.29	< 0.020	< 0.018	< 0.011	< 0.011
c-25-21-1	38.0	8.5	1.08	0.050						
c-25-21-2	37.4	8.3	1.09	0.050						
c-25-21-3	37.2	8.1	1.21	0.051						
c-25-18-1	36.0	7.2	1.25	0.059						
c-25-18-2	35.4	6.7	1.33	0.068						
c-25-18-3	35.7	6.8	1.25	0.062						
c-25-16-1	35.7	4.8	0.65	0.039						
c-25-16-2	36.4	4.2	0.66	0.043						
c-25-16-3	35.8	4.4	0.72	0.043						

B. CALCULATIONS OF TOXIC GAS YIELDS WITH UNCERTAINTIES

Table 10 contains the yields of the combustion products calculated using the data from Table 7 Table 8, and Table 9. The estimated uncertainties reflect the repeatability of the volume fractions in replicate tests, uncertainties in the other terms in the yields calculations, and degree of proximity of the measured values to the background levels.

Table 10. Yields of Combustion Products from Cone Calorimeter Tests.

Gas	Heat Flux (kW/m²)	Flow Rate (L/s)	Initial O₂ %	Bookcase			Sofa			Cable		
CO_2	50	25	21	1.11	±	1.5%	1.41	±	2.2%	1.04	±	8.5%
			18	1.10	±	0.1%	1.46	±	1.8%	1.17	±	3.0%
			16	1.06	±	3.5%	1.41	±	2.3%	1.11	±	6.8%
	50	12.5	21	1.02	±	1.1%	1.36	±	2.6%	1.02	±	2.5%
			16	1.03	±	0.5%	1.38	±	4.4%	1.02	±	2.6%
			14	0.93	±	0.3%	1.39	±	1.3%	0.88	±	5.9%
	25	25	21	1.14	±	1.2%	1.55	±	6.1%	1.13	±	6.4%
			18	0.89	±	4.4%	1.33	±	30%	1.28	±	3.7%
			16	0.85	±	1.4%	1.36	±	51%	0.68	±	5.7%
CO	50	25	21	0.010	±	7.7%	0.027	±	1.9%	0.054	±	4.8%
			18	0.013	±	2.9%	0.035	±	4.2%	0.067	±	4.3%
			16	0.019	±	7.4%	0.044	±	1.4%	0.071	±	11%
	50	12.5	21	0.006	±	2.6%	0.024	±	0.9%	0.050	±	2.6%
			16	0.009	±	9.1%	0.035	±	7.3%	0.061	±	3.0%
			14	0.032	±	0.8%	0.033	±	3.6%	0.054	±	9.9%
	25	25	21	0.019	±	20%	0.024	±	0.5%	0.051	±	1.2%
			18	0.023	±	17%	0.029	±	16%	0.063	±	7.0%
			16	0.027	±	0.9%	0.029	±	15%	0.042	±	6.2%
HCN	50	25	21		<	0.0084	0.0037	±	29%		<	0.014
			18		<	0.0086	0.0077	±	23%		<	0.016
			16		<	0.0082	0.0125	±	15%		<	0.017
	50	12.5	21		<	0.0043	0.0036	±	15%		<	0.0070
			16		<	0.0047	0.0096	±	43%		<	0.0083
			14		<	0.0049	0.0039	±	50%		<	0.0090
HCl	50	25	21		<	0.0065		<	0.0053	0.23	±	14%
			18		<	0.0066		<	0.0055	0.26	±	12%
			16		<	0.0063		<	0.0058	0.30	±	15%
	50	12.5	21		<	0.0033		<	0.0027	0.22	±	15%
			16		<	0.0036		<	0.0032	0.23	±	14%
			14		<	0.0038		<	0.0033	0.28	±	12%
NO	50	25	21		<	0.019		<	0.015		<	0.030
			18		<	0.019		<	0.016		<	0.036
			16		<	0.018		<	0.017		<	0.038
	50	12.5	21		<	0.010		<	0.0078		<	0.016
			16		<	0.010		<	0.0091		<	0.019
			14		<	0.011		<	0.0094		<	0.020

Gas	Heat Flux (kW/m^2)	Flow Rate (L/s)	Initial O$_2$ %	Bookcase		Sofa		Cable	
NO$_2$	50	25	21	<	0.016	<	0.013	<	0.027
			18	<	0.017	<	0.014	<	0.032
			16	<	0.016	<	0.015	<	0.034
	50	12.5	21	<	0.0084	<	0.0068	<	0.014
			16	<	0.0092	<	0.0080	<	0.016
			14	<	0.010	<	0.0082	<	0.018
Acrolein	50	25	21	<	0.010	<	0.0082	<	0.016
			18	<	0.010	<	0.0085	<	0.019
			16	<	0.010	<	0.0090	<	0.021
	50	12.5	21	<	0.0051	<	0.0042	<	0.0083
			16	<	0.0056	<	0.0049	<	0.010
			14	<	0.0058	<	0.0050	<	0.011
Formaldehyde	50	25	21	<	0.011	<	0.0088	<	0.017
			18	<	0.011	<	0.0091	<	0.021
			16	<	0.010	<	0.010	<	0.022
	50	12.5	21	<	0.0055	<	0.0044	<	0.0089
			16	<	0.0060	<	0.0052	<	0.011
			14	<	0.0063	<	0.0054	<	0.011

The values in Table 11 are the values from Table 10 divided by the notional yields from Table 3. Thus the uncertainties are the combined uncertainties from those two tables.

Table 11. Fractions of Notional Yields.

Gas	Heat Flux (kW/m²)	Flow Rate (L/s)	Initial O₂ %	Bookcase			Sofa			Cable		
CO₂	50	25	21	0.65	±	2.5%	0.72	±	3.2%	0.49	±	9.5%
			18	0.64	±	1.1%	0.75	±	2.8%	0.55	±	4.0%
			16	0.62	±	4.5%	0.72	±	3.3%	0.53	±	7.8%
	50	12.5	21	0.59	±	1.9%	0.70	±	3.6%	0.48	±	3.5%
			16	0.60	±	0.8%	0.71	±	5.4%	0.48	±	3.6%
			14	0.54	±	0.5%	0.71	±	2.3%	0.42	±	6.9%
	25	25	21	0.66	±	2.1%	0.79	±	7.1%	0.53	±	7.4%
			18	0.52	±	7.5%	0.68	±	31%	0.61	±	4.7%
			16	0.50	±	2.4%	0.70	±	52%	0.32	±	6.7%
CO	50	25	21	0.009	±	8.7%	0.022	±	5.9%	0.041	±	5.8%
			18	0.012	±	8.7%	0.028	±	8.2%	0.051	±	5.3%
			16	0.018	±	8.7%	0.036	±	5.4%	0.054	±	12%
	50	12.5	21	0.006	±	8.7%	0.020	±	4.9%	0.037	±	3.6%
			16	0.008	±	8.7%	0.028	±	11%	0.046	±	4.0%
			14	0.029	±	8.7%	0.027	±	7.6%	0.040	±	11%
	25	25	21	0.017	±	8.7%	0.020	±	4.5%	0.038	±	2.2%
			18	0.021	±	8.7%	0.023	±	20%	0.047	±	8.0%
			16	0.025	±	8.7%	0.024	±	19%	0.031	±	7.2%
HCN	50	25	21	<		0.15	0.019	±	33%	<		0.34
			18	<		0.15	0.040	±	27%	<		0.41
			16	<		0.14	0.065	±	19%	<		0.43
	50	12.5	21	<		0.076	0.019	±	19%	<		0.18
			16	<		0.082	0.050	±	47%	<		0.21
			14	<		0.087	0.020	±	54%	<		0.23
HCl	50	25	21	<		2.5	<		0.77	0.69	±	14%
			18	<		2.6	<		0.80	0.78	±	13%
			16	<		2.4	<		0.85	0.91	±	16%
	50	12.5	21	<		1.3	<		0.39	0.67	±	15%
			16	<		1.4	<		0.46	0.71	±	14%
			14	<		1.5	<		0.47	0.86	±	12%

V. DISCUSSION

A. OVERALL TEST VALUES

The principal outcome of this series of tests is a well-documented set of combustion product yields. This includes the numerical values themselves, the apparatus conditions under which they were obtained, the uncertainty in their calculated values, and the repeatability of the tests.

Next most important is a determination of the extent to which the toxic gas yields are affected by variations in the test protocol that are reasonable in light of possible variations in combustion conditions in fires involving the intact products.

Third, it is important to evaluate the quality of the derived knowledge in the context of its intended use. The yield information would be used with a computational fire model (zone or CFD) to generate the time-dependent environment generated by a fire. Equations such as those in ISO 13571[9] would then be used to assess whether the combination of occupancy design, contained combustibles, and occupant/responder characteristics lead to the desired level of life safety.

The documentation of the yields has been provided in the earlier sections. The following examines the context and quality of the results.

B. SPECIMEN PERFORMANCE AND TEST REPEATABILITY

1. CO_2 and CO

Changing the oxygen concentration had little effect on CO_2 yields. In some cases at the lowest oxygen concentrations there was a measurable reduction. In the case of the sofa materials, at the lower heat flux, reducing the oxygen concentration increased the variability in the CO_2 yield considerably. This was the result of the combined effect of low heat flux and low oxygen failing to sustain combustion of the specimen, resulting in some degree of non-flaming pyrolysis in which the specimen mass failed to oxidize to either CO or CO_2, but instead escaped as unreacted hydrocarbon (which we did not measure).

On the other hand, CO yields at 50 kW/m^2 and 25 L/s were affected by altering the available oxygen. As can be seen in Figure 4 to Figure 6, decreasing the available oxygen to 16 % increased the CO yields—about a twofold increase for the bookcase, a 60 % increase for the sofa, and a 30 % increase for the cable. When the oxygen volume percent was lower than 16 %, this resulted in a very large increase in the CO yield, a factor of 4 for the bookcase, or a slight decrease (less than 10 %) from the maximum yield at 16 % oxygen, for the sofa and cable.

It can also be seen in these figures that reducing the air flow to 12.5 L/s resulted in consistently lower CO yields. There are several possible explanations: for example, the lower flow may have increased the effluent residence time near the hot cone, allowing more time to oxidize, or the slower flow may have allowed more steady combustion, whereas a more fluctuating flow at the higher flow has the potential to choke off and quench pockets that have not fully oxidized.

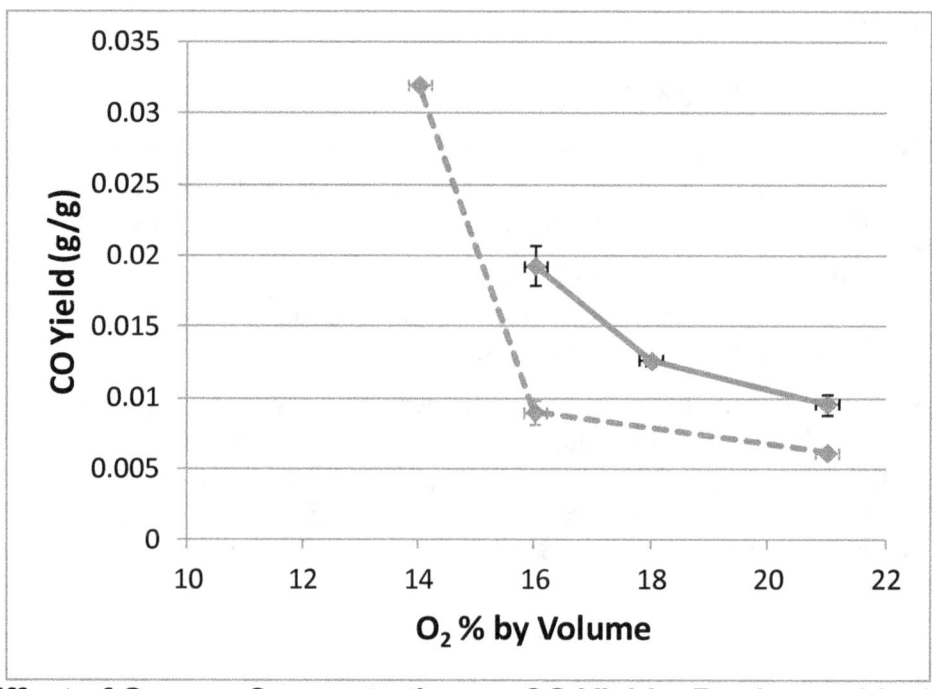

Figure 4. Effect of Oxygen Concentration on CO Yields, Bookcase (dashed line = reduced flow). Error bars are ± the standard deviation of yields from multiple runs.

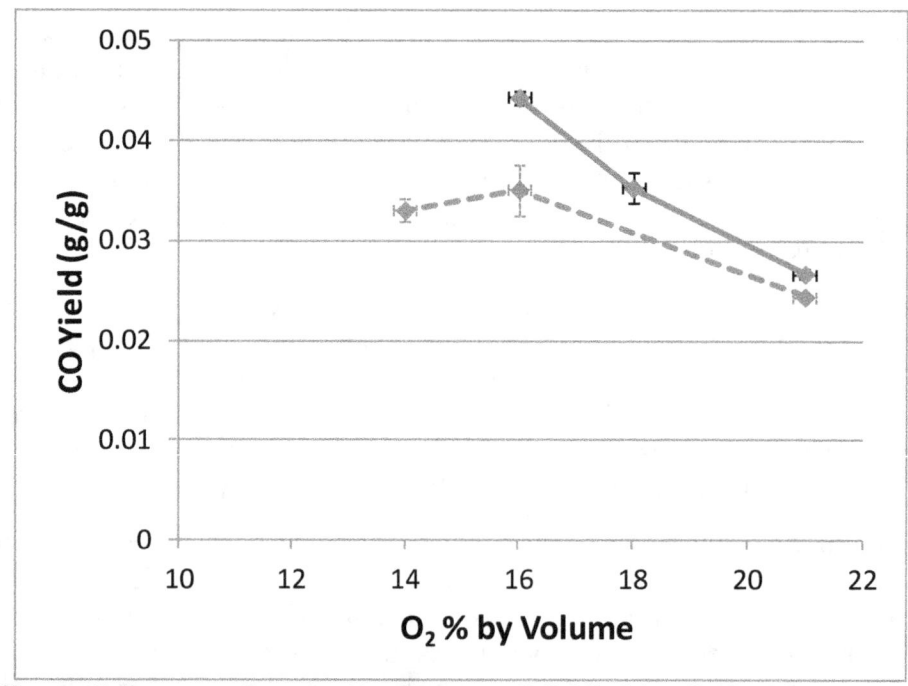

Figure 5. Effect of Oxygen Concentration on CO Yields, Sofa (dashed line = reduced flow). Error bars are ± the standard deviation of yields from multiple runs.

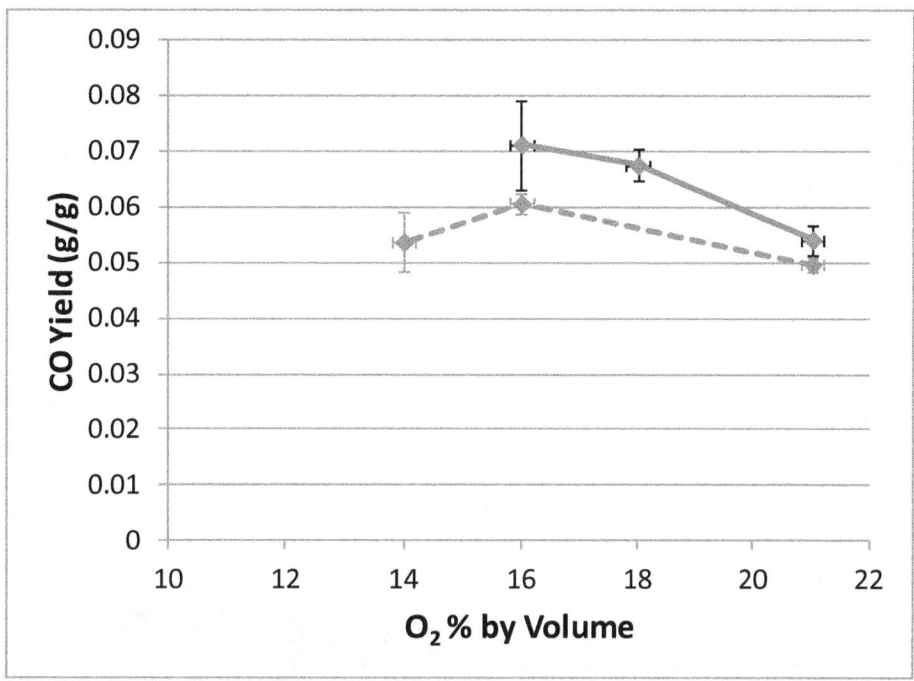

Figure 6: Effect of Oxygen Concentration on CO Yields, Cable (dashed line = reduced flow). Error bars are ± the standard deviation of yields from multiple runs.

2. HCl and HCN

HCN was detected via FTIR and was only observed in the sofa material tests. Figure 7 shows the calculated volume fraction from the FTIR spectra for each of 3 experiments, all conducted at 50 kW/m^2, 12.5 L/s flow, and 16 % oxygen by volume. The three superimposed plots have been time shifted so that the time of ignition is normalized. One of the runs (circles) ultimately resulted in a calculated yield roughly half of the other two, resulting in a large variability of 33 % (see Table 10). If this run is discarded, then the reported yield of HCN at this condition should be 0.0114 with an uncertainty of 0.2%.

It is also worth noting that the total time for combustion extended to about 120 s on the scale in Figure 7. Other combustion products continue to be observed in this time period. In other words, the bulk of the HCN is produced early in the combustion process.

Figure 8 shows the results, as yields, for the generation of HCN at different oxygen volume fractions. As with the CO, there is linear increase in HCN yields as the oxygen volume fraction decreases from 0.21 to 0.16, but at the lowest oxygen volume fraction the yield of HCN is quite low. Another important consideration for Figure 8 is that if the outlier run is excluded, then the results from the 12.5 L/s experiments are much closer to the 25 L/s ones, although still slightly lower. The outlier was included because we feel it is representative of the variability of this measurement.

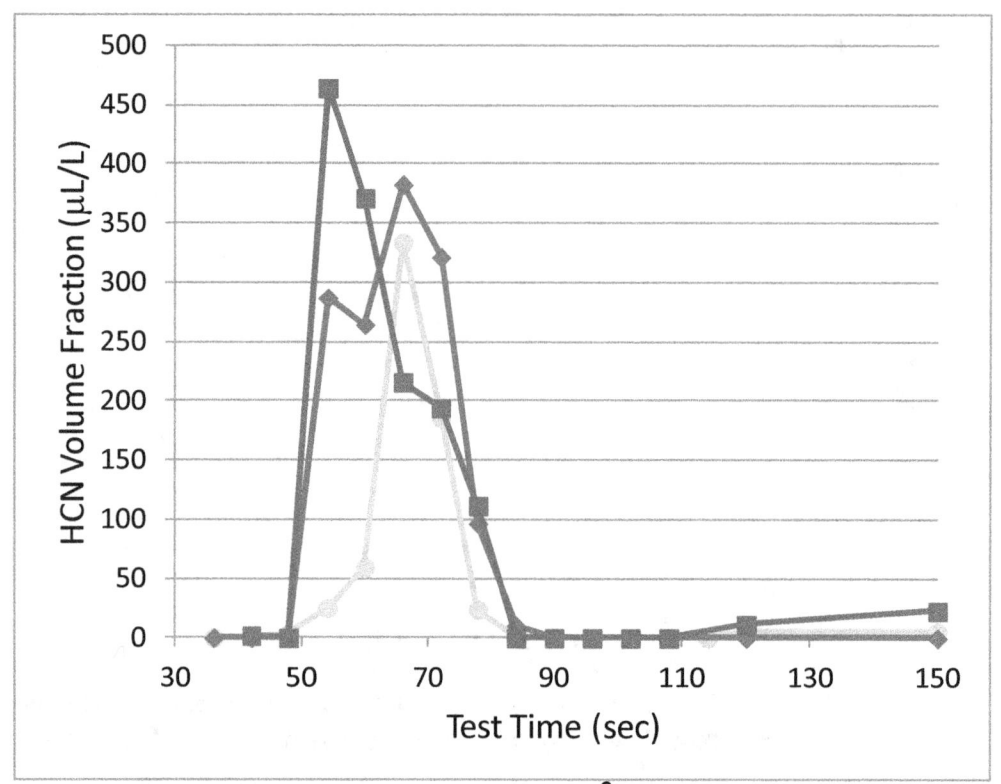

Figure 7. HCN Volume Fractions, Sofa, 50 kW/m², 12.5 L/s, 16 % O₂ by Volume. Uncertainty is 10 % of the Reported HCN Volume Fraction

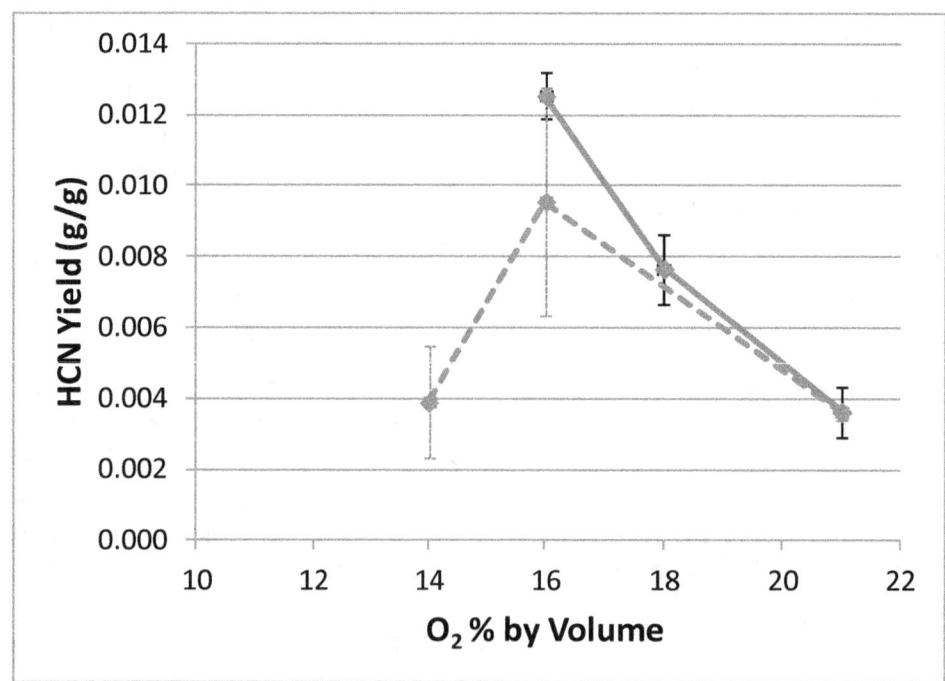

Figure 8. Effect of Oxygen Concentration on HCN Yields, Sofa. Error bars are ± the standard deviation of yields from multiple runs

HCl was detected via FTIR and was only observed in the cable tests. Figure 9 shows the calculated volume fraction from the FTIR spectra for each of 3 experiments, all conducted at 50 kW/m^2, 12.5 L/s flow and 16 % oxygen by volume. The three superimposed plots have been time shifted so that the time of ignition is normalized. The 3 runs are quite consistent, and numerically have a variability in yield of 3.8 %. This is despite the fact that there is some variability in the timing of the release of HCl, lasting approximately 20 s longer for one of the experiments (diamonds in the figure). In other words, the measured volume fraction at any one time is more variable than the integrated quantity.

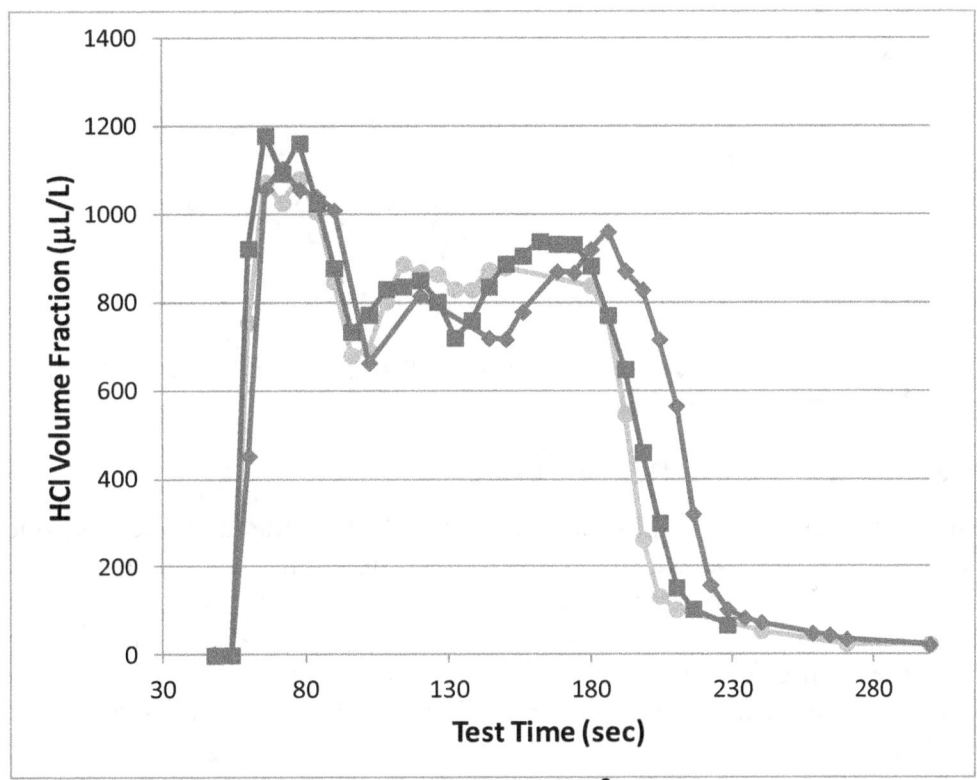

Figure 9. HCl Volume Fractions, Cable, 50 kW/m^2, 12.5 L/s, 16 % O$_2$ by Volume. Uncertainty is 10 % of the Reported Volume Fraction.

Figure 10 shows the effect of oxygen concentration on the yields of HCl from the cable materials. The yields are somewhat higher at lower oxygen concentrations, but this may be accounted for by a slightly lower total mass lost—if disproportionately more of the carbon remains in the char. At the lowest oxygen volume fraction, 0.14, the yield of HCl increases by about 20 %.

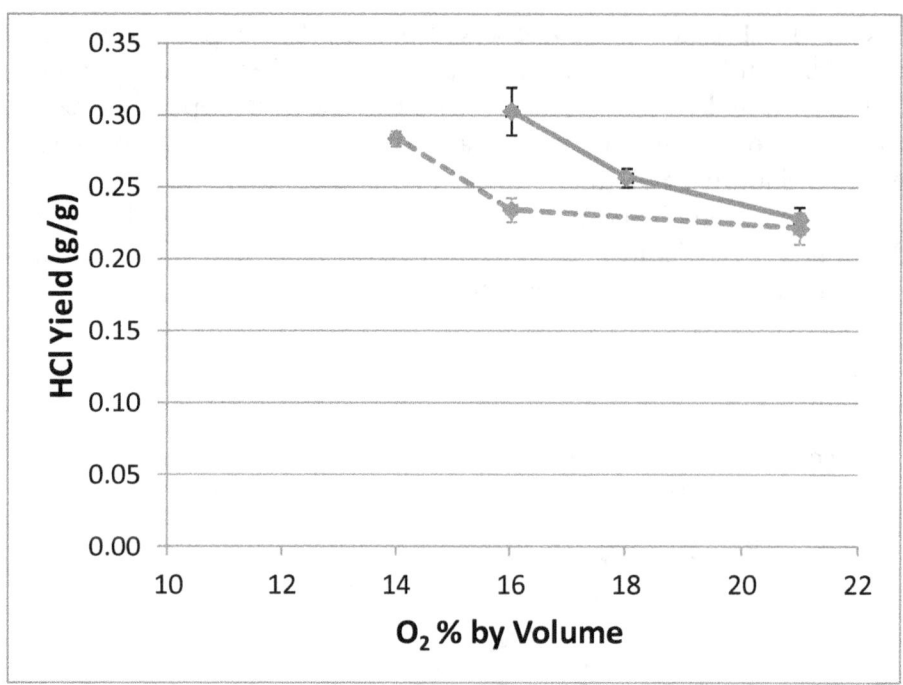

Figure 10. Effect of Oxygen Concentration on Yields of HCl, Cables. Error bars are ± the standard deviation of yields from multiple runs

C. MEASURED VS. NOTIONAL VALUES

During sustained and complete combustion, the yield of CO_2 should approach its notional values, since CO_2 is a marker for combusted carbon. In these experiments, the sofa material came out best, achieving over 70 % of the notional yield. The bookcase fared slightly worse, achieving only 65 % of the notional yield for CO_2, and the cable had the lowest at around half. The relatively lower yields of CO_2 from the cables is an expected result of flame inhibition due to the chlorine. These trends are consistent with our qualitative observation of, for example, the thickness of the smoke produced.

The yields of CO from all the specimens ranged from about 0.01 to 0.06, which corresponds to 1 % to 5 % of the notional values. These values are consistent with relatively fuel-lean combustion[34] and are an order of magnitude lower than the CO yields expected of postflashover fires.[14]

The yields of HCl from the cable specimens approach their notional values. The deficit may reflect scavenging by the calcium carbonate filler in the cable jacket or wall loss in the duct or sampling line.

The detection limited yields of HCl from the sofa specimens was 80 % of the notional yield at the higher flow rate, and 40% of the notional yield at the lower flow rate. The detection limited yield of HCl from the bookcase was 2.5 and 1.5 times the notional yield, for the high and low flows respectively. Chlorine is present at well under one percent by mass in these two products. Such small quantities are not capable of producing toxicologically significant yields of HCl, suggesting that they can be either ruled out by elemental analysis as unimportant, or that limits of detection are not a barrier to measuring significant quantities of HCl in this apparatus.

Yields of HCN from the sofa are between 2 % and 6 % of their notional values. The highest relative yield is found at 16 % oxygen by volume. Despite the lower oxygen, these flames appear relatively fuel-lean, and the nitrogen may appear as molecular nitrogen or nitrogen dioxides.

D. SPECIES SAMPLING AND MEASUREMENT

1. Species Measurement Using FTIR Spectroscopy

FTIR spectroscopic analysis of combustion products has become fairly common in fire research laboratories. However, that does not mean that its use is straightforward. The data from a recent round robin involving FTIR measurement of toxic combustion products from a standardized apparatus showed interlaboratory variations of up to an order of magnitude. There are documents under development in ISO TC92 SC1 and SC3 to standardize the implementation.

We were able to obtain usable information using this technique. There are a number of lessons emerging from this test series that can provide useful input to these efforts, such as the following:

- The application of FTIR spectroscopy to fire testing requires the constant attention of an experienced professional at a level well beyond the demands of the more traditional fire test instrumentation.

- In this work we used a longer path length cell than in previous work. However, the increased sensitivity was offset by the instantaneous nature of the measurements (as opposed to averaging over dozens of data points) and the fact that the gases did not accumulate as they do in the closed chamber tests such as NFPA 269 and NFPA 270 (toxicity test method and smoke density chamber respectively). The CO concentrations measured here are approximately one-tenth of those measured in the closed-chamber apparatus. However, non-linear effects resulted in a 25 % deficit compared to the NDIR measurements of CO at the highest concentrations.

- The cell volume remained small and provided good time resolution. For example, the measured HCl concentrations go from zero to the maximum in two data intervals (see Figure 9).

- The lack of a filter did not adversely affect the measurements. Compared to previous work with other apparatus, the smoke measured here was relatively dilute. The cell only required cleaning a few times over the course of all of the experiments reported here. The internal mirrors were easily cleaned with ethanol and deionized water. The KBr windows were replaced once.

- A heated sample line (as recommended in the SAFIR report[35] and ISO 19702[36]) enabled near-quantitative collection of HCl, a compound that is generally regarded as difficult to determine.

E. IMPORTANCE OF UNDETECTED GASES

The equations in ISO 13571 include provision for additional gases to be included in estimating the time available for escape or refuge from a fire: HBr, HF, SO_2, NO_2, acrolein (C_3H_4O) and formaldehyde (H_2CO). There was no Br, F, or S in any of the products examined in this project, so the first three of these gases were not expected. The presence of the latter three was not detected, thus establishing the upper limits of their presence at the volume fractions listed in Table 2.

To put the potential contributions of the sensory irritant gases (HCl, NO_2, acrolein, and formaldehyde) in context, we use the equations in ISO 13571 for calculating the Fractional Effective Dose (FED) for the narcotic gases, CO_2 and CO, and the Fractional Effective Concentration (FEC) for the four sensory irritant gases.

The FED equation is:

$$FED = \left[\sum_{t1}^{t2} \frac{\varphi_{CO}}{35\,000} \Delta t + \sum_{t1}^{t2} \frac{\varphi_{HCN}^{2.36}}{1.2 \cdot 10^6} \Delta t \right] \exp\left[\frac{\varphi_{CO_2}}{5} \right],$$

where Δt is the exposure interval in minutes.

The FEC of the four irritant gases were estimated from their volume fractions and the incapacitating levels in ISO 13571 (F_{HCl}, etc.). The results are compiled in Table 12.

The FEC equation in ISO 13571 is:

$$FEC = \frac{\varphi_{HCl}}{F_{HCl}} + \frac{\varphi_{HBr}}{F_{HBr}} + \frac{\varphi_{HF}}{F_{HF}} + \frac{\varphi_{SO_2}}{F_{SO_2}} + \frac{\varphi_{NO_2}}{F_{NO_2}} + \frac{\varphi_{acrolein}}{F_{acrolein}} + \frac{\varphi_{formaldehyde}}{F_{formaldehyde}} + \sum \frac{\varphi_{irritant}}{F_{C_i}}$$

Table 12. Limits of importance of undetected toxins

		Volume fraction (μL/L)				FEC Contribution			
		HCl	NO$_2$	C$_3$H$_4$O	H$_2$CO	HCl	NO$_2$	C$_3$H$_4$O	H$_2$CO
Incapacitating Level		1000	250	30	250				
Bookcase	<	20	< 40	< 20	< 40	< 0.02	< 0.16	< 0.66	< 0.16
Sofa	<	20	< 40	< 20	< 40	< 0.02	< 0.16	< 0.66	< 0.16
Cable		1370	< 40	< 20	< 40	0.58	< 0.07	< 0.28	< 0.07
	to	730	< 40	< 20	< 40	0.43	< 0.09	< 0.39	< 0.09

It stands out that the FEC contribution from acrolein is as much as two-thirds of an incapacitating level. This is because (a) the limit of detection is close to the listed incapacitating level and (b) the incapacitating level is very low. While there is agreement among experts that this value of 30 μL/L is reasonable, there are data that suggest strongly that this is unnecessarily conservative. Kaplan and co-workers exposed individual baboons to various concentrations of acrolein in air.[37] At the end of 5 min, each baboon was given a signal and could perform an action that led to escape from the test chamber. The baboons exposed at up to 500 μL/L escaped and survived. Those exposed to higher levels escaped, but died later. These data suggest that people should be able to accommodate a nearly instantaneous exposure to, e.g., at least 300 μL/L without becoming incapacitated. If we increase the incapacitating level of acrolein to 250 μL/L, the relative contribution to FEC become those in Table 13. In this case the relative importance of acrolein falls below 0.1 when other irritants are detected.

Table 13. Limits of Importance of Undetected Toxicants

	Volume fraction (µL/L)				FEC Contribution			
	HCl	NO$_2$	C$_3$H$_4$O	H$_2$CO	HCl	NO$_2$	C$_3$H$_4$O	H$_2$CO
Incapacitating Level	1000	250	250	250				
Bookcase	< 20	< 40	< 20	< 40	< 0.05	< 0.38	< 0.19	< 0.38
Sofa	< 20	< 40	< 20	< 40	< 0.05	< 0.38	< 0.19	< 0.38
Cable	1370	< 40	< 20	< 40	0.77	< 0.09	< 0.05	< 0.09
to	730	< 40	< 20	< 40	0.65	< 0.14	< 0.07	< 0.14

Where specimens like the bookcase and sofa produce narcotic gases, CO and HCN, and irritants are below the limits of detection, the narcotic gases are the primary contributors to incapacitation. Assuming that the irritant gases are present in just under the limits of detection, in order for them to reach an incapacitating concentration, the relative concentration of CO would be incapacitating in a few minutes; the relative concentration of HCN would be incapacitating in under a minute.

This page intentionally left blank

VI. CONCLUSIONS

This paper reports toxic gas yield data for specimens cut from three complex combustibles: a bookcase, a sofa, and residential electrical power cable. The physical fire model used was the cone calorimeter from ISO 5660-1 / ASTM E 1354. This apparatus allows the use of a test specimen that approximates the geometry and radiant exposure that might be experienced by the intact combustible in a well-ventilated flaming fire. In addition to performing the tests as prescribed in the standards, this work added an enclosure and a gas supply capable of reduced oxygen, in order to better approximate conditions in an underventilated fire.

For the standard test procedure:

- The CO_2 yields were very repeatable and represented between half and 80 % of the carbon in the specimens. All specimens left a black residue that continued to pyrolyze after the flaming halted. These residues were presumably carbon-enriched, accounting for the yields being below the notional yields.

- The CO yields were also very repeatable, with the exception of the bookcase at low heat flux and the higher oxygen concentrations.

- The HCN yields were below the limit of detection for the bookcase and cable specimens. The HCN yields from the sofa were 3 to 10 times the limit of detection and had a variation in repeatability between 5 % and 40 % of the reported yield depending on the test conditions.

- The HCl yields were below the limit of detection for the bookcase and sofa specimens. For the cables they were well above the limit of detection and accounted for 70 % of the Cl in the specimens.

- None of the other irritant gases appeared in concentrations above their limits of detection.

Regarding the variation in test conditions, we conclude:

- An incident heat flux of 25 kW/m^2 does not provide any information beyond that which is found at 50 kW/m^2. In fact it leads to increased variability from test to test because of slow ignition or early extinction.

- The trend of toxic gas yields increasing with decreasing initial oxygen concentration holds uniformly across gases and specimen types over the oxygen volume fraction range of 0.21 to 0.16. When the oxygen volume fraction is 0.14, the results are unpredictable for both gases and items burned. Therefore, we don't recommend testing at this level? With the exception of CO from the bookcase, the majority of the CO yields were lower at 0.14 than at 0.16.

- Reducing the total flow to 12.5 L/s reduces the limit of detection of the gases by a factor of two, and allows the achievement of lower oxygen concentrations. However, yields measured at the lower flow were consistently lower than at the higher flow.

Calculation of the contributions of the gases to incapacitation of people who might be exposed to these environments showed:

- Incapacitation from the bookcase material effluent would be primarily from CO.

39

- Incapacitation from the sofa material effluent would be from a combination of CO and HCN.
- Incapacitation from the cable effluent would be initially from HCl; the related yield of CO would become incapacitating after approximately15 minutes.

If the CO yield were at the expected postflashover value of 0.2,

- Incapacitation from the bookcase material effluent would be primarily from CO.
- Incapacitation from the sofa material effluent would be primarily from CO.
- Incapacitation from the cable material effluent would be initially from HCl; a 0.2 yield of CO would become incapacitating after a few minutes.

VII. ACKNOWLEDGEMENTS

The authors express their appreciation to Randy Shields for his assistance in performing the tests.

References

1. Phillips, W.G.B., Beller, D.K., and Fahy, R.F., "Computer Simulation for Fire Protection Engineering," Chapter 5-9 in *SFPE Handbook of Fire Protection Engineering*, 4th Edition, P.J. DiNenno *et al.*, eds., NFPA International, Quincy, MA, 2008.

2. http://www.nist.gov/el/fire_research/cfast.cfm.

3. Peacock, R.D., Jones, W.W., and Reneke, P.A., "CFAST—Consolidated Model of Fire Growth and Smoke Transport (Version 6) Software Development and Evaluation Guide", NIST Special Publication 1086, National Institute of Standards and Technology, Gaithersburg, MD, 187 pages (2008).

4. Peacock, R.D., Jones, W.W., and Bukowski, R.W., "Verification of a Model of Fire and Smoke Transport," *Fire Safety Journal* **21**, 89-129 (1993).

5. http://fire.nist.gov/fds.

6. "Standard Test Method for Heat and Visible Smoke Release Rates for Materials and Products Using an Oxygen Consumption Calorimeter," ASTM E1354-04a, ASTM International, West Conshohocken, PA, 2004.

7. See, *e.g.*, "Standard for the Flammability (Open Flame) of Mattress Sets," 16 CFR Part 1633, *Federal Register* **70** (50), 13471-13523, 2006.

8. http://www.iso.org/iso/iso_catalogue/catalogue_tc/catalogue_tc_browse.htm?commid=50540&published=on&development=on.

9. "Life-threatening Components of Fire – Guidelines for the Estimation of Time to Compromised Tenability in Fires," ISO 13571, International Standards Organization, Geneva, 2012.

10. "Guidelines for Assessing the Fire Threat to People," ISO 19706, International Standards Organization, Geneva, 2011.

11. Gann, R.G., Babrauskas, V., Peacock, R.D., and Hall, Jr., J.R., "Fire Conditions for Smoke Toxicity Measurement," *Fire and Materials* **18**, 193-199, (1994).

12. "Guidance for Comparison of Toxic Gas Data between Different Physical Fire Models and Scales," ISO 29903, International Standards Organization, Geneva, 2012.

13. "Standard Test Method for Developing Toxic Potency Data for Use in Fire Hazard Modeling," NFPA 269, NFPA International, Quincy, MA, 2012.

14. "Standard Test Method for Measuring Smoke Toxicity for Use in Fire Hazard Analysis," ASTM E1678-10, ASTM International, West Conshohocken, PA, 2010.

15. (a) Gann, R.G., Averill, J.D., Nyden, M.R., Johnsson, E.L., and Peacock, R.D., "Smoke Component Yields from Room-scale Fire Tests," NIST Technical Note 1453, National Institute of Standards and Technology, Gaithersburg, MD, 159 pages (2003).

    (b) Gann, R.G., Averill, J.D., Nyden, M.R., Johnsson, E.L., and Peacock, R.D., "Fire Effluent Component Yields from Room-scale Fire Tests," *Fire and Materials* **34**, 285-314, (2010), DOI: 10.1002/fam.1024.

16. Ohlemiller, T.J., and Villa, K., "Furniture Flammability: An Investigation of the California Bulletin 133 Test. Part II: Characterization of the Ignition Source and a Comparable Gas Burner," NISTIR 4348, National Institute of Standards and Technology, Gaithersburg, MD, 1990.

17. Gann, R.G., Marsh, N.D., Averill, J.A., and Nyden, M.R., "Smoke Component Yields from Bench-scale Fire Tests: 1. NFPA 269/ASTM E 1678," NIST Technical Note 1760, National Institute of Standards and Technology, Gaithersburg, MD, 57 pages, (2013).

18. Kaplan, H.L., Grand, A.F., and Hartzell, G.E., *Combustion Toxicology: Principles and Test Methods*, Technology Publishing Co., Lancaster, PA, 1983.

19. "Guidance for Evaluating the Validity of Physical Fire Models for Obtaining Fire Effluent Toxicity Data for Use in Fire Hazard and Risk Assessment – Part 2: Evaluation of Individual Physical Fire Models," ISO/TR 16312-2, International Standards Organization, Geneva, 2007.

20. "Controlled Equivalence Ratio Method for the Determination of the Hazardous Components of Fire Effluent," ISO/TS 19700, International Standards Organization, 2010.

21. "Reaction-to-fire Tests – Heat Release, Smoke Production, and Mass Loss Rates – Part 1: Heat Release (Cone Calorimeter Method)," ISO 5660-1, International Standards Organization, Geneva, 2002.

22. "Plastics — Smoke Generation — Part 2: Determination of Optical Density by a Single-chamber Test, ISO 5659-2, International standards Organization, Geneva, 1994.

23. Marsh, N.D., Gann, R.G., and Nyden, M.R., "Smoke Component Yields from Bench-scale Fire Tests: 2. ISO 19700 Controlled Equivalence Ratio Tube Furnace" NIST Technical Note 1761, National Institute of Standards and Technology, Gaithersburg, MD, 45 pages (2013).

24. Marsh, N.D., Gann, R.G., and Nyden, M.R., "Observations on the Generation of Toxic Products in the NFPA/ISO Smoke Density Chamber," Proceedings of the 12th International Conference on Fire Science & Engineering, (Interflam 2010), Nottingham UK, (2010).

25. Mulholland, G., Janssens, M., Yusa, S., Twilley, W., and Babrauskas, V."The effect of oxygen concentration on CO and Smoke Produced by Flames," Proceedings of the Third International Symposium on Fire Safety Science, Edinburgh UK, pp. 585-594, (1991). doi:10.3801/IAFSS.FSS.3-585

26. Babrauskas, V., Twilley, W.H., Janssens, M., and Yusa, S., "A cone calorimeter for controlled-atmosphere studies." *Fire and Materials* **16**, 37-43, (1992).

27. Hshieh, F.Y., Motto, S.E., Hirsch, D.B., and Beeson, H.D., "Flammability Testing Using a Controlled-atmosphere Cone Calorimeter," Proceedings of the 18th International Conference on Fire Safety Science, Millbrae CA, p. 999 (1993).

28. Gomez, C., Janssens, M., and Zalkin, A. "Using the Cone Calorimeter for Quantifying Toxic Potency" Proceedings of the 12th International Fire Science & Engineering Conference (Interflam 10), Nottingham UK, (2010).

29. Gomez, C., Janssens, M., and Zalkin, A., "Measuring the Yields of Toxic Gases from Materials during Different Stages of Fire Development" In *Fire and Materials* 12 international conference San Francisco, CA (2011).

30. Guillaume, E., Marquis, D.M., and Chivas-Joly, C., "Usage of Controlled-atmosphere Cone Calorimeter to Provide Input Data for Toxicity Modeling," *in Fire and Materials, 12 International Conference,* San Francisco, CA (2011).

31. Haaland, D.M.; Easterling, R.G.; and Vopicka, D.A., "Multivariate Least-Squares Methods Applied to the Quantitative Spectral Analysis of Multicomponent Samples" *Applied Spectroscopy* **39**, 73-84, (1985). doi:10.1366/0003702854249376

32. *Gas Phase Infrared Spectral Standards, Revision B*, Midac Corp.; Irvine, CA (1999).

33. Speitel, L.C., "Fourier Transform Infrared Analysis of Combustion Gases," Federal Aviation Administration Report DOT/FAA/AR-01/88, 2001.

34. Pitts, W.M., "The Global Equivalence Ratio Concept and the Formations Mechanisms of Carbon Monoxide in Fires," *Progress in Energy and Combustion Science* **21**, 197-237, (1995).

35. "Smoke Gas Analysis by Fourier Transform Infrared Analysis,: The SAFIR Project," VTT Research Note, Technical Research Centre of Finland, 81 Pages, 1999.

36. "Toxicity testing of fire effluents -- Guidance for analysis of gases and vapours in fire effluents using FTIR gas analysis," ISO 19702, International Standards Organization, Geneva, 2006.

37. Kaplan, H.L., Grand, A.F., Switzer, W.G., Mitchell, D.S., Rogers, W.R., and Hartzell, G.E., "Effects of Combustion Gases on Escape Performance of the Baboon and the Rat," *Journal of Fire Sciences* **3**, 228-244, (1985).

www.ingramcontent.com/pod-product-compliance
Lightning Source LLC
Chambersburg PA
CBHW081905170526
45167CB00007B/3164